NEXT GENERATION HEALTHCARE SYSTEMS USING SOFT COMPUTING TECHNIQUES

This book presents soft computing techniques and applications used in healthcare systems, along with the latest advancements. Written as a guide for assessing the roles that these techniques play, the book will also highlight implementation strategies, list problem-solving solutions, and pave the way for future research endeavors in smart and next-generation healthcare systems.

This book provides applications of soft computing techniques related to healthcare systems and can be used as a reference guide for assessing the roles that various techniques, such as machine learning, fuzzy logic, and statical mathematics, play in the advancements of smart healthcare systems. The book presents the basics as well as advanced concepts to help beginners, as well as industry professionals, get up to speed on the latest developments in healthcare systems. The book examines descriptive, predictive, and social network techniques and discusses analytical tools and the important role they play in finding solutions to problems in healthcare systems. A framework of robust and novel healthcare techniques is highlighted, as well as implementation strategies and a setup for future research endeavors.

Healthcare Systems Using Soft Computing Techniques is a valuable resource for researchers and postgraduate students in healthcare systems engineering, computer science, information technology, and applied mathematics. The book introduces beginners to—and at the same time brings industry professionals up to speed with—the important role soft computing techniques play in smart healthcare systems.

Artificial Intelligence in Smart Healthcare Systems
Series Editors: Vishal Jain and Jyotir Moy Chatterjee

The progress of the healthcare sector is incremental as it learns from associations between data over time through the application of suitable big data and IoT frameworks and patterns. Many healthcare service providers are employing IoT-enabled devices for monitoring patient healthcare, but their diagnosis and prescriptions are instance-specific only. However, these IoT-enabled healthcare devices are generating volumes of data (Big-IoT Data) that can be analyzed for more accurate diagnosis and prescriptions. A major challenge in the above realm is the effective and accurate learning of unstructured clinical data through the application of precise algorithms. Incorrect input data leading to erroneous outputs with false positives shall be intolerable in healthcare as patients' lives are at stake. This new book series addresses various aspects of how smart healthcare can be used to detect and analyze diseases, the underlying methodologies, and related security concerns. Healthcare is a multidisciplinary field that involves a range of factors like the financial system, social factors, health technologies, and organizational structures that affect the healthcare provided to individuals, families, institutions, organizations, and populations. The goals of healthcare services include patient safety, timeliness, effectiveness, efficiency, and equity. Smart healthcare consists of m-health, e-health, electronic resource management, smart and intelligent home services, and medical devices. The Internet of Things (IoT) is a system comprising real-world things that interact and communicate with each other via networking technologies. The wide range of potential applications of IoT includes healthcare services. IoT-enabled healthcare technologies are suitable for remote health monitoring, including rehabilitation, assisted ambient living, etc. In turn, healthcare analytics can be applied to the data gathered from different areas to improve healthcare at a minimum expense.

This new book series is designed to be a first choice reference at university libraries, academic institutions, research and development centres, information technology centres, and any institutions interested in using, designing, modelling, and analysing intelligent healthcare services. Successful application of deep learning frameworks to enable meaningful, cost-effective personalized healthcare services is the primary aim of the healthcare industry in the present scenario. However, realizing this goal requires effective understanding, application, and amalgamation of IoT, Big Data, and several other computing technologies to deploy such systems in an effective manner. This series shall help clarify the understanding of certain key mechanisms and technologies helpful in realizing such systems.

Designing Intelligent Healthcare Systems, Products, and Services Using Disruptive Technologies and Health Informatics
Teena Bagga, Kamal Upreti, Nishant Kumar, Amirul Hasan Ansari, and Danish Nadeem

Next Generation Healthcare Systems Using Soft Computing Techniques
Rekh Ram Janghel, Rohit Raja, Korhan Cengiz, and Hiral Raja

NEXT GENERATION HEALTHCARE SYSTEMS USING SOFT COMPUTING TECHNIQUES

Edited by
Rekh Ram Janghel, Rohit Raja,
Korhan Cengiz, and Hiral Raja

CRC Press
Taylor & Francis Group
Boca Raton London New York

CRC Press is an imprint of the
Taylor & Francis Group, an **informa** business

First edition published 2023
by CRC Press
6000 Broken Sound Parkway NW, Suite 300, Boca Raton, FL 33487-2742

and by CRC Press
4 Park Square, Milton Park, Abingdon, Oxon, OX14 4RN

CRC Press is an imprint of Taylor & Francis Group, LLC

© 2023 Taylor & Francis Group, LLC

Reasonable efforts have been made to publish reliable data and information, but the author and publisher cannot assume responsibility for the validity of all materials or the consequences of their use. The authors and publishers have attempted to trace the copyright holders of all material reproduced in this publication and apologize to copyright holders if permission to publish in this form has not been obtained. If any copyright material has not been acknowledged please write and let us know so we may rectify in any future reprint.

Except as permitted under U.S. Copyright Law, no part of this book may be reprinted, reproduced, transmitted, or utilized in any form by any electronic, mechanical, or other means, now known or hereafter invented, including photocopying, microfilming, and recording, or in any information storage or retrieval system, without written permission from the publishers.

For permission to photocopy or use material electronically from this work, access www.copyright.com or contact the Copyright Clearance Center, Inc. (CCC), 222 Rosewood Drive, Danvers, MA 01923, 978-750-8400. For works that are not available on CCC please contact mpkbookspermissions@tandf.co.uk

Trademark notice: Product or corporate names may be trademarks or registered trademarks and are used only for identification and explanation without intent to infringe.

ISBN: 978-1-032-10797-4 (hbk)
ISBN: 978-1-032-10799-8 (pbk)
ISBN: 978-1-003-21709-1 (ebk)

DOI: 10.1201/9781003217091

Typeset in Times
by MPS Limited, Dehradun

Contents

Preface vii
Editors ix
Contributors xi

Chapter 1 Computational Intelligence for Healthcare 1

                Abhilasha Chaudhuri and Tirath Prasad Sahu

Chapter 2 Analysis of Recurrent Neural Network and Convolution Neural Network Techniques in Blood Cell Classification 15

                Tatwadarshi P. Nagarhalli

Chapter 3 Evaluating the Effectiveness of the Convolution Neural Network in Detecting Brain Tumors 29

                Tatwadarshi P. Nagarhalli, Sneha Mhatre, Ashwini Save, and Sanket Patil

Chapter 4 Implementation of Machine Learning in Color Perception and Psychology: A Review 41

                Anusruti Mitra, Dipannita Basu, and Ahona Ghosh

Chapter 5 Early Recognition of Dynamic Sleeping Patterns Associated with Rapid Eyeball Movement Sleep Behavior Disorder of Apnea Patients Using Neural Network Techniques 55

                Prateek Pratyasha and Saurabh Gupta

Chapter 6 Smart Attendance cum Health Check-up Machine for Students/Villagers/Company Employees 71

                Pranjal Patel, Shriram Sharma, Pritesh Sutrakar, Hemant Kumar, Devender Pal Singh, and Menka Yadav

Chapter 7 Oral Histopathological Photomicrograph Classification Using Deep Learning 85

                Rajashekhargouda C. Patil and P. K. Mahesh

Chapter 8 Prediction of Stage of Alzheimer's Disease DenseNet Deep Learning Model 105

                Yogesh Kumar Rathore and Rekh Ram Janghel

Chapter 9	An Insight of Deep Learning Applications in the Healthcare Industry	123
	Deevesh Chaudhary, Prakash Chandra Sharma, Akhilesh Kumar Sharma, and Rajesh Tiwari	
Chapter 10	Expand Patient Care with AWS Cloud for Remote Medical Monitoring	137
	Parul Dubey and Pushkar Dubey	
Chapter 11	Privacy and Security Solution in Wireless Sensor Network for IoT in Healthcare System	149
	Rajesh Tiwari, Deevesh Chaudhary, Tarun Dhar Diwan, and Prakash Chandra Sharma	
Chapter 12	An Epileptic Seizure Detection and Classification Based on Machine Learning Techniques	169
	Lokesh Singh, Rekh Ram Janghel, and Satya Prakash Sahu	
Chapter 13	Analysis of Coronary Artery Disease Using Various Machine Learning Techniques	187
	Saroj Kumar Pandey, Rekh Ram Janghel, Shubham Shukla, and Yogadhar Pandey	
Index		209

Preface

This book aims to provide relevant information on applications of soft computing techniques using various biomedical databases related to healthcare systems. There are various issues and challenges in the healthcare system that are resolved by the use of soft computing techniques. Due to the advancement of technologies, this book will be a helpful guide for categories and assess the major role of various soft computing technologies, such as machine learning and deep learning for diagnosis and prediction of disease. The book will explore the various learning model applications in soft computing techniques, and it covers the deep learning, machine learning, artificial neural network, fuzzy logic, evolutionary computing, and statistical techniques. It presents basic and advanced concepts to help beginners and industry professionals get up to speed on the latest developments in soft computing and healthcare systems. It incorporates the latest methodologies and challenges facing soft computing; examines descriptive, predictive, and social network techniques; and discusses analytics tools and their role in providing effective solutions for science and technology. The primary users for the book include researchers, academicians, postgraduate students, specialists, and practitioners. This book will

- Review the state-of-the-art in healthcare processing models, methods, techniques etc.
- Review learning methods in medical health and description.
- Highlight new techniques and applications in biomedical imaging with practical implementation.
- Tackle existing and emerging image challenges and opportunities in the healthcare field.
- Promote mutual understanding of researchers in different disciplines, as well as networking.
- Facilitate future research development and collaborations.

We express our appreciation to all of the contributing authors who helped us tremendously with their contributions, time, critical thoughts, and suggestions to put together this peer-reviewed edited volume. The editors are also thankful to CRC Press and their team members for the opportunity to publish this volume. Lastly, we thank our family members for their love, support, encouragement, and patience during the entire period of this work.

Editors

Rekh Ram Janghel received a PhD in information technology (2013) from ABV-IIITM Gwalior. His main research interest includes face recognition and identification, digital image processing, signal processing, and networking. Presently, he is working as assistant professor in NIT Raipur. He has academic and research experience in various areas of CSE and IT. He has filed and published successfully 11 patents. He has received many invitations to be a guest at IEEE conferences. Dr. Janghel has published 50 research papers in various international/national journals (including IEEE, Springer etc.) and proceedings of the reputed international/national conferences (including Springer and IEEE). He has been nominated for the board of editors/reviewers of many peer-reviewed and refereed journals.

Rohit Raja received a PhD in computer science and engineering from CVRAMAN University in 2016. His main research interests include face recognition and identification, digital image processing, signal processing, and networking. Presently he is working as associate professor in the IT Department, Guru Ghasidas Vishwavidyalaya, Bilaspur (CG), India. He has authored several journal and conference papers. He has good academics & research experience in various areas of CSE and IT. He has filed and published successfully 12 patents. He has received an invitation to be a guest at the IEEE Conference. He has published 77 research papers in various international/national journals (including IEEE and Springer) and proceedings of the reputed international/national conferences (including Springer and IEEE). He has been nominated for the board of editors/reviewers of many peer-reviewed and refereed journals (including IEEE, Springer).

Korhan Cengiz, PhD, SMIEEE was born in Edirne, Turkey, in 1986. He received BS degrees in electronics and communication engineering from Kocaeli University, Turkey, and business administration from Anadolu University, Turkey, in 2008 and 2009, respectively. He earned his MS degree in electronics and communication engineering from Namik Kemal University, Turkey, in 2011, and a PhD degree in electronics engineering from Kadir Has University, Turkey, in 2016. Since 2018, he has been an assistant professor with the Electrical-Electronics Engineering Department, Trakya University, Turkey. He is the author of over 40 articles, including in *IEEE Internet of Things Journal, IEEE Access, Expert Systems with Applications and Knowledge Based Systems,* three book chapters, two international patents, and one book in Turkish. His research interests include wireless sensor networks, wireless communications, statistical signal processing, indoor positioning systems, power electronics, and 5G. He is associate editor of *Interdisciplinary Sciences: Computational Life Sciences,* Springer; handling editor of *Microprocessors and Microsystems,* Elsevier; associate editor of *IET Electronics Letters,* IET Networks; and editor of *AEÜ – International Journal of Electronics and Communications,* Elsevier. He has guest editorial positions in *IEEE Internet of*

Things Magazine and *CMC-Computers, Materials & Continua*. He serves in several reviewer positions for *IEEE Internet of Things Journal, IEEE Sensors Journal,* and *IEEE Access*. He serves in several book editorial positions for Springer, Elsevier, Wiley, and CRC. He presented ten plus keynote talks in reputed IEEE and Springer Conferences about WSNs, IoT, and 5G. He is a senior member, IEEE, since August 2020. Dr. Cengiz's awards and honors include the Tubitak Priority Areas Ph.D. Scholarship, the Kadir Has University PhD Student Scholarship, best presentation award in ICAT 2016 Conference, and best paper award in ICAT 2018 Conference.

Hiral Raja is working as associate professor in the Mathematics Department at Dr. C V Raman University Bilaspur. She received her PhD in mathematics in 2017 from C. V. Raman University India. Her main research includes image processing, embedded system, artificial intelligence, and sensor signaling. She was successfully granted two national and two international patents. She has published approximately ten research papers in national and international journals, including Scopus, IEEE, and Springer.

Contributors

Dipannita Basu
Department of Information Technology
Maulana Abul Kalam Azad University of Technology
West Bengal, India

Deevesh Chaudhary
Department of Information Technology
Manipal University
Jaipur, India

Abhilasha Chaudhuri
Department of Computer Science and Engineering
School of Engineering
O. P. Jindal University
Raigarh, Chhattisgarh, India

Tarun Dhar Diwan
Chhattisgarh Swami Vivekanand Technical University
Bhilai, India

Parul Dubey
Department of Information Technology
Shri Shankaracharya Institute of Professional Management and Technology
Raipur, India

Pushkar Dubey
Department of Management
Pandit Sundarlal Sharma (Open) University
Chhattisgarh, India

Ahona Ghosh
Department of Computer Science and Engineering
Maulana Abul Kalam Azad University of Technology
West Bengal, India

Saurabh Gupta
Department of Biomedical Engineering
National Institute of Technology
Raipur, Chhattisgarh, India

Rekh Ram Janghel
Department of Information Technology
National Institute of Technology
Raipur, Chhattisgarh, India

Hemant Kumar
Electronics and Communication Engineering
Malaviya National Institute of Technology
Jaipur, India

Sneha Mhatre
Vidyavardhini's College of Engineering and Technology
Vasai, Mumbai, India

Anusruti Mitra
Department of Information Technology
Maulana Abul Kalam Azad University of Technology
West Bengal, India

Tatwadarshi P. Nagarhalli
Vidyavardhini's College of Engineering and Technology
Vasai, Mumbai, India

P. K. Mahesh
Department of Electronics and Communication Engineering
Academy for Technical and Management Excellence College of Engineering
Mysore, India

Saroj Kumar Pandey
Department of Computer Engineering and Applications
GLA University
Mathura (U.P.), India

Yogadhar Pandey
Department of Computer Science & Engineering
Technocrats Institute of Technology-Excellence
Bhopal (M.P.), India

Pranjal Patel
Electronics and Communication Engineering
Malaviya National Institute of Technology
Jaipur, India

Rajashekhargouda C. Patil
Department of Electronics and Communication Engineering
Don Bosco Institute of Technology
Bengaluru, India

Sanket Patil
Vidyavardhini's College of Engineering and Technology
Vasai, Mumbai, India

Prateek Pratyasha
Department of Biomedical Engineering
National Institute of Technology
Raipur, Chhattisgarh, India

Yogesh Kumar Rathore
Shri Shankaracharya Institute of Professional Management and Technology
Raipur, Chhattisgarh, India

Satya Prakash Sahu
National Institute of Technology
Raipur, Chhattisgarh, India

Tirath Prasad Sahu
Department of Information Technology
National Institute of Technology
Raipur, Chhattisgarh, India

Ashwini Save
Computer Engineering
VIVA Institute of Technology
Virar, Mumbai, India

Akhilesh Kumar Sharma
Department of Information Technology
Manipal University
Jaipur, India

Prakash Chandra Sharma
Department of Information Technology
Manipal University
Jaipur, India

Shriram Sharma
Electronics and Communication Engineering
Malaviya National Institute of Technology
Jaipur, India

Shubham Shukla
Department of Electronics & Communication Engineering
Krishna Institute of Engineering & Technology-Ghaziabad
Ghaziabad (U.P.), India

Lokesh Singh
Department of Computer Science and Engineering
Alliance University
Banglore, Karnataka, India

Devender Pal Singh
Electronics and Communication Engineering
Malaviya National Institute of Technology
Jaipur, India

Pritesh Sutrakar
Electronics and Communication
  Engineering
Malaviya National Institute of
  Technology
Jaipur, India

Menka Yadav
Electronics and Communication
  Engineering
Malaviya National Institute of
  Technology
Jaipur, India

Rajesh Tiwari
Department of Computer Science &
  Engineering
CMR Engineering College
Hyderabad, India

1 Computational Intelligence for Healthcare

Abhilasha Chaudhuri
Department of Computer Science and Engineering, School of Engineering, O. P. Jindal University, Raigarh, Chhattisgarh, India

Tirath Prasad Sahu
Department of Information Technology, National Institute of Technology, Raipur, Chhattisgarh, India

CONTENTS

1.1 Introduction 2
    1.1.1 Artificial Neural Network 3
    1.1.2 Restricted Boltzman Machines 3
    1.1.3 Support Vector Machines 3
    1.1.4 Evolutionary Algorithms 3
    1.1.5 Fuzzy Systems 4
    1.1.6 Swarm Intelligence 4
1.2 Issues and Challenges 4
    1.2.1 Data Inconsistency, Inaccuracy, and Missing Values 5
    1.2.2 Imbalanced Data 5
    1.2.3 Data Collection Cost 5
    1.2.4 Huge Data Volume 5
    1.2.5 Ethical and Privacy Issues 6
1.3 Feature Engineering 6
    1.3.1 Feature Extraction 6
    1.3.2 Feature Selection 6
        1.3.2.1 Filter Method 6
        1.3.2.2 Wrapper Method 7
        1.3.2.3 Embedded Method 7
    1.3.3 Feature Weighting 7
    1.3.4 Introduction to Gene Expression Dataset 8
    1.3.5 Challenges in Gene Expression Data 8
    1.3.6 Feature Selection and Classification of Gene Expression Data Using Binary Jaya Algorithm 8
        1.3.6.1 Binary Jaya Algorithm 9

DOI: 10.1201/9781003217091-1

     1.3.6.2 Use of Feature Selection with Binary Jaya Algorithm 9
     1.3.6.3 Result and Discussion 10
1.4 Available Resources 11
1.5 Conclusion 12
References 12

1.1 INTRODUCTION

When it comes to healthcare, we're no exception to the rule of a society fascinated with statistics. Increasing amounts of information are being generated (Huang et al. 2015). Without a way to extract knowledge and information from all of this data, it's pointless (Chen et al., 2006). Data science and computational intelligence meet healthcare in this area. This chapter explores how artificial intelligence can be used in the healthcare industry. By organising, processing, and analysing a huge and diverse set of data, computational intelligence approaches try to extract knowledge and insights. Recursive Boltzman machines, support vector machines, fuzzy systems, the evolutionary algorithm, and swarm intelligence are some of the methodologies used by data scientists. In this chapter, we'll go through these methods in detail. Here, we'll take a look at how data science can be used in the healthcare industry. Aside from the money, lives are at stake in the healthcare industry. Approaches and models used must therefore be exceedingly precise and effective. Accuracy of the machine learning model is dependent on the quality of the training data. Data errors in the healthcare industry are extremely prevalent, such as missing data and unbalanced data. This chapter focuses on data-related issues and challenges. In this chapter, we'll examine feature engineering as a solution to these kinds of problems. Healthcare is projected to be significantly impacted by genetics and genomics, which will disclose new dimensions such as personalised medicine, disease prediction, and genetic engineering. We'll look at a real-world example of something like this in action in this section. In order to better understand the principles presented in this chapter, some open-source datasets and tools are also explained. Methods of Computer-Assisted Reasoning are widely used in the field of healthcare nowadays. The important Computational Intelligence Techniques as shown in Figure 1.1 are discussed here.

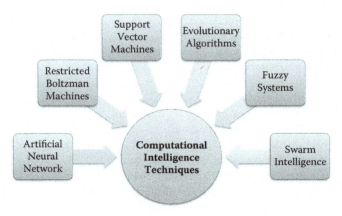

FIGURE 1.1 Computational intelligence techniques.

1.1.1 ARTIFICIAL NEURAL NETWORK

The goal of an artificial neural network (ANN) is to emulate the brain's own neural network. Multiple neurons are arranged in a layered pattern to create an architecture similar to that of a biological neural network using mathematical modelling of neurons. It is important to remember that each cell has a set of input connections (which represent synapses on the cell and its dendrite), an output value (which represents the neuron's firing rate), a bias value (which represents the neuron's internal resting level), and a set of output connections (which represent the neuron's axonal projection). Each of these perspectives on the unit is represented mathematically using real numbers. Consequently, every link has a weight (synaptic strength) that affects the unit's activity level in response to the incoming input. By modifying the weights in accordance with a learning method, we can make an artificial neuron learn (Chandrakar et al. 2021).

1.1.2 RESTRICTED BOLTZMAN MACHINES

An energy-based, undirected generative model known as the RBM (Chandrakar et al. 2021) uses hidden variables to represent a distribution of visible variables. We can ensure that the likelihood term's posterior contribution across hidden variables is almost factorial using the undirected model, which considerably aids inference. Using an energy-based model, you may determine how likely it is that certain variables will occur. Because there are no visible or hidden linkages, it is referred to as "limited."

1.1.3 SUPPORT VECTOR MACHINES

Classification and regression can be accomplished with SVMs, a collection of supervised learning techniques. In the dataset's input space, a linear or nonlinear separation surface is utilised to classify the data. As a typical tool for computational intelligence and data mining, SVM delivers the best performance in real-world applications like text categorization, handwritten character recognition, image classification, bio-sequence analysis, and so on. An implicit transfer of input data onto a much higher dimensional feature space where the data can be linearly separated is achieved using kernels. When a linear decision boundary is drawn, the margin, or shortest distance between training instances and the border, is maximised. A cost is imposed to account for cases that were erroneously classified, and the margin is maximised while the cost is minimised if the mapped data points are linearly inseparable.

1.1.4 EVOLUTIONARY ALGORITHMS

It is possible to use evolutionary algorithms (EAs) to solve search and optimization problems based on evolutionary processes in biological organisms (Chaudhuri and Sahu 2020; Hancer et al. 2018). It takes several generations for natural populations to adapt in accordance with natural selection and "survival of the fittest." If these algorithms are properly encoded, they can "evolve" solutions to real-world problems.

1.1.5 FUZZY SYSTEMS

Certainty, ambiguity, opacity, and inconsistency are all common occurrences in the real world. It is therefore imperative that new logic representations be provided for effective modelling in order to satisfy these ambiguous real-world scenarios. Fuzzy sets and logic are gaining popularity in today's technology landscape because of this. Computational intelligence, clustering, control, data analysis and mining, medical diagnostics, optimization, and pattern classification are just a few of the fields where these technologies have been employed.

1.1.6 SWARM INTELLIGENCE

Swarm intelligence is a type of decentralised intelligence that focuses on the collective behaviour of intelligent agents (Abdel-Basset et al. 2020; Nguyen, Xue, and Zhang 2020). Local interactions among agents frequently lead a global pattern to merge, despite the fact that there is usually no centralised control guiding the behaviour of the agents. The majority of the basic concepts are inspired by natural swarms, such as ant colonies, bird flocking, honeybees, bacteria, and microbes. Swarm models are population based, having a population of potential solutions as the starting point. In order to discover the best solution, these people are then optimised across many rounds using numerous heuristics inspired by insect social behaviour (Pandey et al. 2022).

1.2 ISSUES AND CHALLENGES

Using the methodologies outlined in section 1.2, expert systems trained using computational intelligence techniques have the potential to assist the healthcare business in a variety of ways. However, there are other data concerns and issues that must be solved. Data, as the foundation of data science technology, must be accurate, consistent, and balanced, as well as available in a sufficient amount to train machine learning algorithms. After all, the quality of an algorithm is only as good as the data used to train it.

The concerns and challenges of data in the healthcare sector will be examined in this section. Figure 1.2 depicts the unique concerns and challenges that data scientists encounter when developing expert systems.

FIGURE 1.2 Issues and challenges related to data in computational intelligence techniques.

1.2.1 Data Inconsistency, Inaccuracy, and Missing Values

True, there are a lot of electronic health records on the market. They are, however, not in a format that is ideal for computational intelligence techniques and algorithms. Accessible data must still be processed before computational intelligence technology may be applied to it. As a result of their usage of automated or semi-automatic software to record these inputs, many companies, or even the same organisation, have saved data in various formats over time. There appears to be a discrepancy in the available data. More than that, there are numerous missing values due to a number of issues, such as a lack of information, privacy concerns, the merging incompatible format, and so on. Data entry errors, a lack of competence in the particular field (of the person recording the data in the system), and other factors contribute to the inaccurate data that are shown.

1.2.2 Imbalanced Data

The number of features that characterise distinct qualities and the number of samples or occurrences make up a dataset. The dataset is said to be imbalanced when the ratio between the number of variables (features) and the number of observations is exceptionally large (Chaudhuri and Sahu 2021b). The gene expression dataset collected by microarray experiment is one of the extremely unbalanced datasets. There are tens of thousands of features but just a few hundred cases. Consider the lung cancer gene expression dataset, which has 12,533 attributes but only 181 instances. On these types of datasets, training a machine learning model is extremely tough. Some datasets for text classification, face recognition, and image classification are skewed as well. Data scientists will need to devise new methods for training mathematical models in unbalanced datasets.

1.2.3 Data Collection Cost

Pathological or medical imaging examinations are used to gather data in the healthcare field. Because data scientists demand a database in the format required by them, developing and maintaining one is prohibitively expensive. Because of these restrictions, data scientists have a hard time locating the best datasets to suit their needs.

1.2.4 Huge Data Volume

Large volumes of data are difficult to collect due to high expenses; on the contrary, the healthcare industry has a tremendous amount of data available (Mehta and Pandit 2018), which is difficult to process. The big data processing has attracted many researchers to work in this field. The medical insurance sector, pharmaceutical corporations, medical equipment design and manufacture, and other industries generate large amounts of data.

1.2.5 Ethical and Privacy Issues

Aside from the technical challenges mentioned previously, the healthcare profession must also cope with ethical and privacy concerns. Without the consent of patients, it is unethical to use or distribute their data. In addition, the patients' privacy should be respected at all times. When a new medicine is released, its ethical implications are also taken into account. In medical insurance agencies, ethical and privacy concerns are especially essential. Individuals' genetic databases are likewise incredibly sensitive datasets that should be handled with privacy concerns in mind.

1.3 FEATURE ENGINEERING

Feature engineering techniques try to extract or find meaningful, informative features from the available feature set of the data. These techniques attempt to reduce the computational cost and improve the accuracy and performance of machine learning algorithms. Feature engineering techniques address the issues and challenges faced by the healthcare sector that were discussed in section 1.2. Feature engineering incorporates the extraction, selection, and weighting of features. This section explores these strategies in depth.

1.3.1 Feature Extraction

New features are derived from the dataset's current feature set using feature extraction techniques. There is a linear or non-linear combination of the original features that results in the new features that have been derived. Because the new feature set has fewer features than the original feature set, feature extraction techniques minimise the dataset's dimensionality. It is through this process of feature extraction that machine learning algorithms are made more efficient.

1.3.2 Feature Selection

Methods for picking the most useful subset of features from a large dataset minimise its dimensionality (Chaudhuri and Sahu 2021c). There are no linear or non-linear combinations used to construct the new features; rather, they are simply selected from the original features set and created as a subset. ML models benefit from this subset of features because they are more efficient and perform better because of it. Filtering, wrapping, and embedding are three types of feature selection strategies (Tiwari, et al. 2021). Here, you'll learn about each of these methods.

1.3.2.1 Filter Method

Filter methods pick the feature subset based on the dataset's qualities. They use statistical and information theoretic techniques to examine and recognise the hidden pattern in the dataset. During the feature selection phase, the classifier is not involved. As a result, these methods are thought to be more generic, quick, and cost-effective feature selection methods. Figure 1.3 depicts the feature selection process

Computational Intelligence for Healthcare

FIGURE 1.3 Filter method.

using the filter approach. To evaluate the performance of the feature selection approach, the optimal feature subset is obtained, and then a classifier is applied.

1.3.2.2 Wrapper Method

The classifier is involved in each step of feature selection using wrapper methods for feature selection. They pick a feature subset using search techniques such as forward selection and backward exclusion, and then the classifier evaluates the selected feature subset for accuracy. After then, another feature subset will be tested until the most accurate feature subset is found. This feature subset will then be identified as the final feature subset. Wrapper-based feature selection takes longer than filter approaches because the classifier is involved in each step (Figure 1.4).

1.3.2.3 Embedded Method

The filter and wrapper methods are combined to make an embedded method. They combine the benefits of both filter and wrapper methods in one package. Algorithms with built-in feature selection methods should be used to implement it (Han, Kamber, and Pei 2012) (Figure 1.5).

1.3.3 FEATURE WEIGHTING

By giving weights to characteristics based on their importance, feature weighting approaches reduce data dimensionality (Chaudhuri and Sahu 2021d). Weight values are allocated to more significant qualities. The less significant elements are removed by setting a threshold. Features can be weighted using any real number between 0 and 1, making it more flexible than selecting features (Han, Kamber, and Pei 2012).

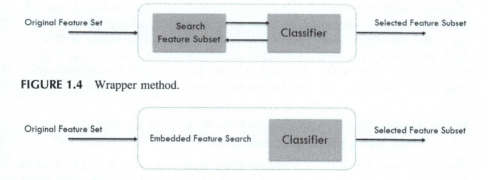

FIGURE 1.4 Wrapper method.

FIGURE 1.5 Embedded method.

A value of 0 indicates that no features are selected, whereas a value of 1 indicates that features are selected. This is the sole way to indicate whether or not features are selected. As a result, feature selection can be thought of as a subset of weighting features in which feature weights are restricted to binary values.

> **CASE STUDY USE OF GENE EXPRESSION DATA CLASSIFICATION FOR DISEASE DIAGNOSIS**
>
> Microarray technology, which yields high throughput data, has made gene expression profiling possible (Chaudhuri and Sahu 2021a; Cleofas-Sánchez, Sánchez, and García 2019). This high-throughput data pertains to gene expression levels. The investigation of various diseases at the genetic level is called gene expression profiling. There are several datasets accessible for diseases such as colon cancer, breast camcer, and ovarian cancer, to name a few.

1.3.4 Introduction to Gene Expression Dataset

Microarray gene expression datasets are very high dimensional datasets. Each feature of the dataset is associated with a particular gene. In other words, the expression level of a gene is listed in a particular column of the dataset. Each dataset contains two types of samples, healthy samples and cancerous samples. Since microarray gene expression profiling is very costly, there are fewer samples in a microarray dataset but a large number of features because the number of genes is very high in organisms. This nature of the gene expression dataset poses many challenges for machine learning models. These challenges are discussed in the next subsection.

1.3.5 Challenges in Gene Expression Data

The large number of features (thousands) and small number of samples (hundreds) make the microarray gene expression dataset very imbalanced. It is very problematic to train a ML algorithm on the imbalanced dataset. It also causes the curse of dimensionality phenomenon. Due to this, the generalization capacity of the ML algorithm becomes very poor. Moreover, the interaction among the genes also makes the learning difficult. In order to overcome these problems, we need to reduce the number of features of the gene expression dataset (Baliarsingh, Vipsita, and Dash 2019; Cleofas-Sánchez, Sánchez, and García 2019). For dimensionality reduction, we have used the wrapper-based feature selection technique.

1.3.6 Feature Selection and Classification of Gene Expression Data Using Binary Jaya Algorithm

The binary Jaya optimization technique (Chaudhuri and Sahu 2021c) is used to choose features and classify gene expression data. Consider a gene expression dataset with 7129 genes and 60 samples pertaining to the central nervous system (CNS). It's a binary classification dataset, which means there are just two classes. The link http://csse.szu.edu.cn/sta/zhuzx/Datasets.html can be used to get this dataset.

1.3.6.1 Binary Jaya Algorithm

A simple, effective, and rapid population-based metaheuristic algorithm is Jaya. It does not require any algorithm-specific parameter tuning. The Jaya algorithm ensures that the best solution is found while avoiding the worst. At each iteration of the algorithm, the best and worst solutions from the population are tracked down. Based on the best and worst answers in Equation (1.1), the Jaya algorithm's population solution is modified

$$X_{k,j}^{t+1} = X_{k,j}^t + r1 \times (Best_j - |X_{k,j}^t|) - r2 \times (Worst_j - |X_{k,j}^t|) \qquad (1.1)$$

for a D dimensional dataset and population size N. The population's k^{th} solution is indicated by the symbol X_k. The current value of the jth dimension of the kth solution is represented by $X_{k,j}^t$. $Best_j$ and $Worst_j$ denote the jth dimension of the population's best and worst solution, respectively; $r1$ and $r2$ symbolise the random numbers generated in the range [0, 1].

The Jaya algorithm described till now works fine on the continuous space. However, feature selection is a discrete optimization problem; therefore, we need to use the following transfer function described in Equation (1.2) and Equation (1.3)

$$X_{k,j}^{t+1} = \begin{cases} 0 & \text{if } r \text{ and } < T(X_{k,j}^t) \\ 1 & \text{if } r \text{ and } \geq T(X_{k,j}^t) \end{cases} \qquad (1.2)$$

$$T(X_{k,j}^t) = \frac{1}{1 + e^{-X_{k,j}^t}} \qquad (1.3)$$

to convert the search space from continuous to discrete. Figure 1.6 depicts the overall working of the Jaya algorithm.

1.3.6.2 Use of Feature Selection with Binary Jaya Algorithm

First, the number of features in the dataset will be determined; in this case, there are 7129 features in the dataset. The population size will then be determined; for example, let's say the population size is 30. The result will be a two-dimensional 307×129 matrix. Each column of the matrix represents a characteristic of the dataset, whereas each row represents a particle or candidate solution vector. The solution vector can only have two values: 0 or 1, with 0 indicating that the associated feature is not selected and 1 indicating that it is selected. The population is randomly initialized.

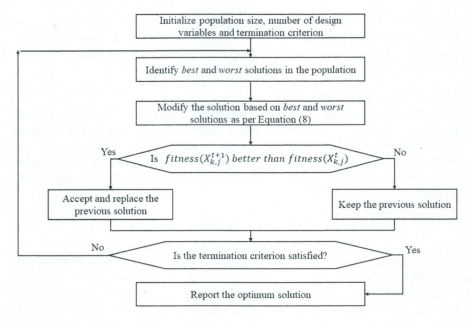

FIGURE 1.6 Flow chart of Jaya algorithm.

The *best* and the *worst* solutions are identified in the polulation based on the fitness of each solution. Fitness is calculated using the classifier. It is nothing but the classification accuracy of the classifier based on the selected feature subset. Then, the population is updated using equation (1.1). At this point of time, the solutions become continuous, which is again converted into binary format with the help of equation (1.2) and (1.3). The algorithm proceeds in this manner until the termination criteria are met.

1.3.6.3 Result and Discussion

We used four classifiers to assess the fitness of each solution: Logistic Regression, Nave Bayes, Decision Trees, and KNN. Binary Jaya was used to choose features for the CNS gene expression dataset, and the results are shown in Table 1.1. Table 1.1

TABLE 1.1

Comparative study of the classification accuracy of several classifiers, both with and without the application of feature selection

Classifier	Accuracy without Feature Selection	Accuracy with Binary Jaya Based Feature Selection
KNN	48.15	74.07
Decision Tree	66.67	81.48
Naïve Bayes	55.55	75.93
Logistic Regression	60.0	72.26

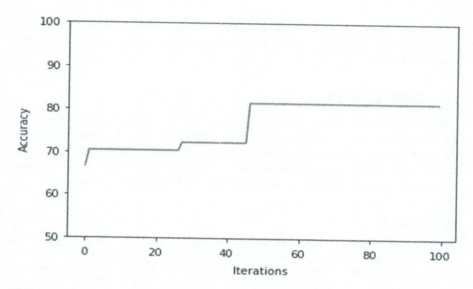

FIGURE 1.7 Convergence curve of decision tree classifier over CNS dataset.

shows the average accuracies of ten algorithm runs. The detailed examination of Table 1.1 reveals that when features are selected, classification accuracy improves dramatically. The reason for this is that the classifier performs better when irrelevant, redundant, and noisy characteristics are removed. As a result, it's a good idea to use a feature selection strategy before starting a classification work. The highest classification accuracy is achieved using the decision tree classifier. The convergence behaviour is depicted in Figure 1.7. The binary-Jaya-based feature selection algorithm was employed to achieve this accuracy.

1.4 AVAILABLE RESOURCES

The following repositories have a large number of datasets relevant to various diseases:

- The University of California Irvine's (UCI) machine learning repository can be accessed at https://archive.ics.uci.edu/ml/datasets.php
- Kaggle is a fantastic resource for finding publically available datasets. https://www.kaggle.com/datasets
- The Indian government makes healthcare datasets available at https://data.gov.in/

Aside from the data, some of the platforms also include free software for the most often used data science approaches.

- https://www.kaggle.com/notebooks
- https://github.com/donnemartin/data-science-ipython-notebooks

To acquire hands-on experience with the concepts addressed in this chapter, readers should explore the public data sources and code linked to data science methodologies.

1.5 CONCLUSION

Computational intelligence techniques are crucial in healthcare and medicine because of the enormous volume and dimensionality of data. In order to improve the performance of machine learning models, feature engineering approaches are essential. Computer-aided healthcare could lead to better patient care, as well as to better working conditions and experiences for healthcare professionals.

REFERENCES

Abdel-Basset, Mohamed, Doaa El-Shahat, Ibrahim El-henawy, Victor Hugo, C. de Albuquerque, and Seyedali Mirjalili. 2020. "A New Fusion of Grey Wolf Optimizer Algorithm with a Two-Phase Mutation for Feature Selection." *Expert Systems with Applications.* 139: 112824. Elsevier Ltd. 10.1016/j.eswa.2019.112824.

Baliarsingh, Santos Kumar, Swati Vipsita, and Bodhisattva Dash. 2019. "A New Optimal Gene Selection Approach for Cancer Classification Using Enhanced Jaya-Based Forest Optimization Algorithm." *Neural Computing and Applications.* 32(12): 8599–8616. Springer London. 10.1007/s00521-019-04355-x.

Chandrakar, Ramakant, Rohit Raja, and Rohit Miri. 2021. "Animal Detection Based on Deep Convolutional Neural Networks with Genetic Segmentation." *Multimed Tools and Applications.* 73(2): 1–14, 10.1007/s11042-021-11290-4.

Chandrakar, Ramakant, Rohit Raja, Rohit Miri, Upasana Sinha, Alok Kumar Kushwaha, and Hiral Raja. 2021. "Enhanced the Moving Object Detection and Object Tracking for Traffic Surveillance Using RBF-FDLNN and CBF Algorithm." *Expert Systems with Applications.* 191(1), ISSN: 0957-4174. 10.1016/j.eswa.2021.116306.

Chaudhuri, Abhilasha, and Tirath Prasad Sahu. 2020. "Feature Selection Using Binary Crow Search Algorithm with Time Varying Flight Length." *Expert Systems with Applications.* 168(1): 114288. Elsevier.

Chaudhuri, Abhilasha, and Tirath Prasad Sahu. 2021a. "A Case Study on Disease Diagnosis Using Gene Expression Data Classification with Feature Selection: Application of Data Science Techniques in Health Care." In Sharaff, Aakanksha and Sinha, G. R. *Data Science and Its Applications.* 239–254. Chapman and Hall/CRC.

Chaudhuri, Abhilasha, and Tirath Prasad Sahu. 2021b. "A Hybrid Feature Selection Method Based on Binary Jaya Algorithm for Micro-Array Data Classification." *Computers and Electrical Engineering.* 90 (January): 106963. Elsevier Ltd. 10.1016/j.compeleceng.2020.106963.

Chaudhuri, Abhilasha, and Tirath Prasad Sahu. 2021c. "Binary Jaya Algorithm Based on Binary Similarity Measure for Feature Selection." *Journal of Ambient Intelligence and Humanized Computing.* 0123456789. Springer Berlin Heidelberg. https://doi.org/10.1007/s12652-021-03226-5.

Chaudhuri, Abhilasha, and Tirath Prasad Sahu. 2021d. "Feature Weighting for Naïve Bayes Using Multi Objective Artificial Bee Colony Algorithm." *International Journal of Computational Science and Engineering.* 24(1): 74–88. 10.1504/IJCSE.2021.113655.

Chen, Hsinchun, Fuller, Sherrilynne S., Friedman, Caro, & Hersh, William (Eds.). (2006). Medical informatics: knowledge management and data mining in biomedicine (Vol. 8). Springer Science & Business Media.

Cleofas-Sánchez, Laura, J. Salvador Sánchez, and Vicente García. 2019. "Gene Selection and Disease Prediction from Gene Expression Data Using a Two-Stage Hetero-Associative Memory." *Progress in Artificial Intelligence*. 8(1): 63–71. Springer Berlin Heidelberg. 10.1007/s13748-018-0148-6.

Han, Jiawei, Micheline Kamber, and Jian Pei. 2012. *Data Mining: Concepts and Techniques*. Morgan Kuffmann, Elsevier. 10.1016/C2009-0-61819-5.

Hancer, Emrah, Bing Xue, Mengjie Zhang, Dervis Karaboga, and Bahriye Akay. 2018. "Pareto Front Feature Selection Based on Artificial Bee Colony Optimization." *Information Sciences* 422: 462–479. Elsevier Inc. 10.1016/j.ins.2017.09.028.

Huang, Tao, Liang Lan, Xuexian Fang, Peng An, Junxia Min, and Fudi Wang. 2015. "Promises and Challenges of Big Data Computing in Health Sciences." *Big Data Research*. 2(1), 2–11. 10.1016/j.bdr.2015.02.002.

Mehta, Nishita, and Anil Pandit. 2018. "Concurrence of Big Data Analytics and Healthcare: A Systematic Review." *International Journal of Medical Informatics*. 114, 57–65. 10.1016/j.ijmedinf.2018.03.013.

Nguyen, Bach Hoai, Bing Xue, and Mengjie Zhang. 2020. "A Survey on Swarm Intelligence Approaches to Feature Selection in Data Mining." *Swarm and Evolutionary Computation* 54 (April 2019): 100663. Elsevier B.V. 10.1016/j.swevo.2020.100663.

Pandey, Sonam, Rohit Miri, G. R. Sinha, and Rohit Raja. 2022. "AFD Filter and E2N2 Classifier for Improving Visualization of Crop Image and Crop Classification in Remote Sensing Image." *International Journal of Remote Sensing*. 10.1080/01431161.2021.2000062.

Tiwari, Laxmikant, Rohit Raja, Vineet Awasthi, Rohit Miri, G. R. Sinha, Monagi H. Alkinani, and Kemal Polat. 2021. "Detection of Lung Nodule and Cancer Using Novel Mask-3 FCM and TWEDLNN Algorithms." *Measurement* 172: 108882. ISSN 0263-2241. 10.1016/j.measurement.2020.108882.

2 Analysis of Recurrent Neural Network and Convolution Neural Network Techniques in Blood Cell Classification

Tatwadarshi P. Nagarhalli
Vidyavardhini's College of Engineering and Technology,
Vasai, Mumbai, India

CONTENTS

2.1 Introduction...15
    2.1.1 Deep Learning Techniques 16
    2.1.2 Medical Imaging/White Blood Cell Classification 17
2.2 Dataset 19
2.3 Analysis of Deep Learning Techniques 21
    2.3.1 Recurrent Neural Network 22
    2.3.2 Convolution Neural Network...22
    2.3.3 Convolution Neural Network Experiment Design and Results 24
        2.3.3.1 Case # 1 24
        2.3.3.2 Case # 2 25
        2.3.3.3 Case # 3 25
2.4 Conclusions..26
References...27

2.1 INTRODUCTION

Machine learning and deep learning have revolutionised many important sectors, including finance, banking, automobile, and medical. Machine learning is an important subfield of artificial intelligence that endeavours to learn from the examples given to it and produce results for completely unknown data, which is not possible for the normal algorithms.

    Machine learning techniques can be broadly classified into two types, supervised and unsupervised, according to the type of input data provided. In supervised learning, the algorithms are provided with training data that contain targets or the expected output as well, whereas in unsupervised learning, that is not the case.

DOI: 10.1201/9781003217091-2

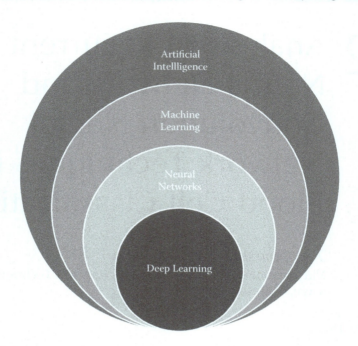

FIGURE 2.1 Artificial intelligence vs machine learning vs deep learning.

Further, the supervised learning techniques can again be classified into two types, regression and classification, according to the type of targets. If the expected output is a finite continuous numerical, then regression algorithms are applied, and if the expected result is a distinct group of variables called classes, classification algorithms are employed.

One of the important classification techniques is the artificial neural network. Artificial neural network tries to mimic the human neural structure by trying to learn in a collaborative manner [1]. A specialisation of an artificial neural network is deep learning techniques and methods.

Figure 2.1 shows the relationship among artificial intelligence, machine learning, and deep learning.

2.1.1 Deep Learning Techniques

One of the main ideas behind the development and usage of deep learning techniques involves extracting deeper and comprehensive intuitions or insights from the given data. Deeper insights are especially required for complex data, like text, images, video, and audio for more accurate predictions. Because of the these deeper insights, the deep learning algorithms have played a very important role in enabling a paradigm shift in the processing of complex information in crucial sectors like medicine, military, and finance, among many others.

An artificial neural network contains only one hidden layer apart from the input layer and the output layers. On the other hand, deep neural networks possess more than one hidden layer. These additional hidden layers enable the deep neural networks to extract deeper insights from the given input [2]. Because of these deeper insights, such neural networks are called deep learning networks.

There are many variants of deep learning techniques that have been proposed, like the deep neural network, autoencoders, the restricted Boltzmann machine, the recurrent neural network, and the convolution neural network [3].

The recurrent neural network and convolution neural network have gained prominence in recent times because of the kind of results produced [4]. In different areas, these two deep learning techniques have proven to be better than other techniques with the kind of insights and level of prediction accuracy produced.

The recurrent neural network and its variants like long short term memory and the gated recurrent unit have been extremely successful while working with text. One major reason for this is the ability of the recurrent neural network to retain contextual information [5]. Contextual information plays a very important part in accurate prediction while working with text. Similar to contextual information, the recurrent neural network can also retain temporal information to some extent. This is the reason why the recurrent neural network has been extensively used and experimented with when working with videos [6].

One of the major drawbacks of the recurrent neural network is that it is a linear process and takes a very long time for training. Considering the fact that most of the deep learning techniques require a large amount of data for training, as the feature extraction process is also done by these techniques, the recurrent neural network requiring a large amount of time is a major hindrance. For better parallelisation of the training data, the convolution neural network was introduced.

The convolution neural network, on the other hand, can work faster and is not a serial process, so it can be parallelised [7]. This quality makes the convolution neural network one of the highly used and experimented techniques that has been adopted in many different areas. This is also the reason why most of the research areas where the convolution neural network has been employed deal with images and very few areas deal with cases where contextual information might be important, like in texts.

As single images can be worked upon independently without any requirement for the knowledge of the previous images, and as different areas of the same image can be processed for insights in parallel, the convolution neural network has become the first choice when working with images. Because of this reason, the convolution neural network technique is often considered and used in computer vision, which is another field of artificial intelligence that exclusively deals with images. And, precisely because of these reasons, the convolution neural network has been employed in the field of medical imaging.

2.1.2 Medical Imaging/White Blood Cell Classification

One of the important sectors where deep learning techniques, and especially the convolution neural network, have been successfully employed and have produced good results is medical imagery. Medical imagery is an important tool in proper diagnosis of diseases. Proper and early diagnosis of diseases is important so that better care can be taken of the patient.

X-rays to sonography, to microscopes, to cameras are all visual aids that help in proper diagnosis of diseases. These visual aids are also the source of imagery on which learning techniques have been employed for an early and quick diagnosis of diseases with minimal human intervention. Because these learning techniques reduce human intervention, the probability of human error has also been reduced.

Deep learning techniques have been extensively experimented with for diagnosis of diseases in medical imagery, for example on magnetic resonance imaging (MRI), X-ray images, and microscopic images, among others [8,9]. One of the areas where these techniques can play an important role is with microscopic images of blood in order to identify blood-related diseases.

One of the types of blood cells is the white blood cell. White blood cells are one of the most important building blocks of the human immune system. White blood cells are the first responders of the human body's immune system, tasked with fighting off infections and foreign invaders [10]. An increase in the count of white blood cells indicates that there is some kind of infection that the body is trying to fight.

There are five type of white blood cells in the human body, monocytes, lymphocytes, neutrophils, eosinophils, and basophils [11]. Each of these blood cells has some unique function. Monocytes are responsible for fighting off chronic infections, whereas lymphocytes are responsible for fighting infections. Eosinophils are responsible for fighting infections caused by parasites, basophils are responsible for fighting allergic reactions, and neutrophils constitute the largest among the white blood cells and search for and fight bacteria and fungus. Figure 2.2 shows the different types of white blood cells found in the human body according to their proportions.

Among these different white blood types Neutrophils constitute for about 60%–70%, Lymphocytes are the net abundant white blood cells found in the human body constituting about 20%–40%, Monocytes constitute about 2%–8%, Eosinophils constitute about 1%–4% and finally Basophils are the white blood cells which are found in least numbers in the human body constituting only around 0.5% [12].

The type of disease can be approximated by identifying the presence and increase in their numbers. That is, by identifying the increase in specific type of these white blood cells a proper diagnosis of a particular disease can be made [13]. This is

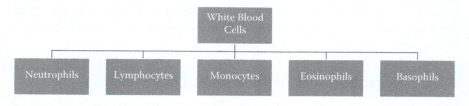

FIGURE 2.2 White blood cell found in the human body.

where deep learning techniques like Recurrent Neural Network and Convolution Neural Network can be applied in identifying the type of white blood cells.

There have been few research works which have implemented and experimented with some of the deep learning techniques in blood cell classification with limited successes. But, for large-scale adaptation of deep learning techniques in medical imagery in general and blood cell classification in particular detailed analysis is required in order to identify the effectiveness of deep learning techniques in this field of research. So, the chapter proposes to conduct a study of different deep learning techniques used for classification of blood cell type and analyse their effectiveness in producing accurate classification results.

2.2 DATASET

For the purpose of classification, the algorithms have to be trained on examples of the data. For this, a dataset is required with a large number of examples. Especially in deep learning techniques like the recurrent neural network and the convolution neural network, which performs feature extraction by itself, a very large number of trainings is required.

The most frequently used image dataset for white blood cell classification is the Blood Cell Images dataset from Kaggle [14]. The dataset contains images of four out of the five white blood cell types. The dataset does not contain examples of Basophils. As Basophils constitutes only around 0.5% of the total white blood cells in the human body. It can be argued that, for the purposes of classification, it is not a major drawback to not have examples of Basophils.

The dataset contains close to 10,000 images of the four white blood cells for training purposes. Each of the four types of white blood cells under consideration has close to around 2500 examples. Table 2.1 shows the exact number of training images for each type of white blood cells present in the dataset.

Figures 2.3, 2.4, 2.5, and 2.6 show sample images of eosinophils, lymphocytes, monocytes, and neutrophils, respectively, from the training data.

From the sample images, it can be understood that the task of image classification becomes difficult as all the microscopic images of the blood cells look very similar to each other. Even though deep learning is preferred when working with complex data like images, the problem of images being similar is amplified when

TABLE 2.1

Number of training images of blood cells

Sl. No.	Name of the White Blood Cells	Number of Training Images
1.	Eosinophils	2497
2.	Lymphocytes	2483
3.	Monocytes	2478
4.	Neutrophils	2499

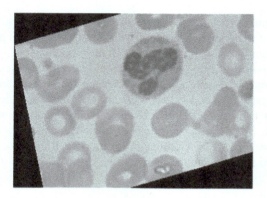

FIGURE 2.3 Eosinophils.

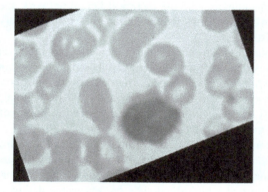

FIGURE 2.4 Lymphocytes.

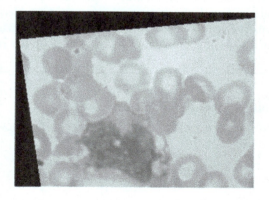

FIGURE 2.5 Monocytes.

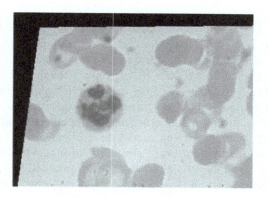

FIGURE 2.6 Neutrophils.

TABLE 2.2
Number of testing images of blood cells

Sl. No.	Name of the White Blood Cells	Number of Testing Images
1.	Eosinophils	623
2.	Lymphocytes	620
3.	Monocytes	620
4.	Neutrophils	624

using deep learning techniques as the network tries to identify the features by itself. Because of this reason, the feature extraction performed by the deep learning techniques is not very precise; this leads to lower accuracy levels.

Another challenge while performing blood cell classification is the number of training examples. As feature extraction is done by the algorithms themselves in deep learning, a large amount of training data is expected. The said dataset under consideration, which has close to 2500 examples each for the four types of white blood cells, cannot be completely adequate; it is just acceptable.

The dataset contains a total of around 2500 images, with each of the four blood cell types having around 600 images. Table 2.2 shows the exact number of testing images present in the dataset.

2.3 ANALYSIS OF DEEP LEARNING TECHNIQUES

There are many different deep learning algorithms and techniques that are employed for different purposes, including the deep neural network, autoencoders, the restricted Boltzmann machine, the recurrent neural network, and the convolution neural network. Among the mentioned deep learning algorithms, the recurrent neural network and the convolution neural network are the most widely used; hence, analysis of these two algorithms will be carried out.

2.3.1 RECURRENT NEURAL NETWORK

Recurrent neural network is the class of deep learning techniques that can handle temporal information. The recurrent neural network is able to do this because of the presence of the back propagation through time network present in the network [15]. This property of the recurrent neural network makes it very useful for the technique to hold on to contextual information as well. Due to these reasons, the recurrent neural network is a popular technique used for text analysis. But, the presence of this network also makes the recurrent neural network very slow, and the temporal information also gets blurred over a longer period. In order to solve the loss of contextual information, variants of recurrent neural network have been proposed like long short term memory and gated recurrent unit.

During the survey of related works for blood cell classification, it was observed that there was hardly any work that employed the recurrent neural network for blood cell classification. As a matter of fact, there were hardly any works that employed the recurrent neural network for image classification. The most that has been done is that researchers have combined the recurrent neural network with the convolution neural network for image classification and specifically blood cell classification [16,17].

That does not mean that the recurrent neural network cannot be used for image classification. In order to test the effectiveness of the recurrent neural network for blood cell classification, a vanilla recurrent neural was implemented for blood cell classification of the dataset mentioned in the previous section. The findings are as follows,

i. The algorithm took a lot of time for training; comparatively it took almost double the amount of time required for the convolution neural network.
ii. Even the time taken for testing was comparatively twice as much.
iii. The accuracy score for the vanilla recurrent neural network for blood cell classification for the said dataset was 11.37%.

From these findings it can said that for image classification in general and blood cell classification in particular the recurrent neural network is not effective, and it isn't that very useful.

2.3.2 CONVOLUTION NEURAL NETWORK

Most of the research on blood cell classification using the deep learning technique uses the convolution neural network. And, most of the research work that has made use of the convolution neural network has claimed to achieve an accuracy score in excess of 90% [18,19]. As already mentioned, one of the main reasons for this is, the basic working of the convolution neural network is more suitable and in a way designed for image analysis.

Some of the important components in a convolution neural network are the convolution layer, the pooling layer, and the dense network. Of these three components, the convolution and the pooling layers have been essentially designed for an image, for its processing in a nonlinear manner.

The convolution layer is essentially a feature extraction component of the convolution neural network, and the pooling layers reinforces the extracted features by, generally, maximising or highlighting it. The dense layer is a fully connected neural network layer used for classification purposes.

At a theoretical level, the application seems very simple, but the practical implementation is very complex, which involves many matrix multiplications. With the availability of many sophisticated packages like TensorFlow, Keras, and TFLearn, it can be argued that the implementation of the complex convolution neural network is simple. Having said that, the real research and challenge is to decide on an architecture of a convolution neural network for a particular problem at hand.

The number of convolution layers, pooling layers, and dense layers can be different and can be modified according to the problem; this is part of the research. Apart from this, the positioning of these layers is also a matter of research. That is, we can have a network architecture where all the convolution layers have been used first, then the pooling layers have been added, or we can have an architecture where it is all mixed-up. These are the different parameters that have to be decided.

Apart from the mentioned parameters, there are many hyper parameters that have to be considered as well. For instance, the size of the input image, the filter size, the stride for the filter, the type of activation function, whether we want the filter size to be the same or smaller, among others, are the hyper parameters that need to be factored in.

Similarly, even for the pooling layer, it has to be first decided what kind of pooling layer is required among the maximum, minimum, and average pooling. Once that has been decided, the next task is to fix the size of the window and the stride. Even for the dense network, the number of output neurons has to be decided.

Another hyper parameter that has to be decided is the number of epochs. Since a deep learning algorithm requires a significantly large training dataset, which is difficult to procure, the same training set is taken for fine tuning other hyper parameters. The number of times the model is expected to work through the entire dataset is the number of epochs.

Apart from these parameters and hyper parameters, another important thing that has to be decided is the type of optimizer. An optimizer is a very important component for any deep learning algorithm. Optimizers are methods and techniques used for altering the values of the weights of the filters in the convolution layer, the weights of the windows in pooling layers, and the learning rates. There are many different optimizers, like the gradient descent, stochastic gradient descent, RMSProp, Adam, and Adaboost.

There are another two factors that complicate the matter. Every time a convolution neural network is trained, it is not a compulsion that the same set of images are taken in a particular order. Generally, even the shuffle option is set so that between epochs the images are shuffled. Otherwise, there is high possibility that the model over-fits, which will lead to very low testing accuracy. Also, each time a convolution neural network is trained, the filter weights and the window weights are taken at random. This also has a significant impact on the overall accuracy of the model.

Because of all the reasons mentioned, it is very difficult to replicate the deep learning network results in general and the convolution neural network in particular. Additionally, it is also true that many research papers only talk about the architecture of the convolution neural network and do not give details about the hyper parameters, which essentially determine the accuracy of the proposed system.

2.3.3 Convolution Neural Network Experiment Design and Results

For the purpose of experimentation and in order to gauge the effectiveness of convolution neural network in white blood cell classification, a few architectures and hyper parameters were tested.

2.3.3.1 Case # 1

The architecture taken was a simple one, with two convolution layers and two max pooling layers applied one after another in alternating manner, that is, one convolution layer then a max pooling layer then the same repeated. The number of filters applied to the first convolution layer was 32 with size 3 × 3, and for the second convolution layer, it was 64 with size 2 × 2. For both the convolution layers, Rectified Linear Unit (ReLU) was the activation function. Then, two dense layers were applied. The number of output neurons for the first dense layer was 256, and for the last layer, it was four, as the number of classes in which the classification has to be done is four. The input size of the image was taken as 150×150. And, the optimizer taken was RMSProp.

Multiple models were created, and the models were tested for a different number of epochs. Five models were created, with the number of epochs as 10, 20, 30, 40, and 50. The result obtained has been shown in Figure 2.7.

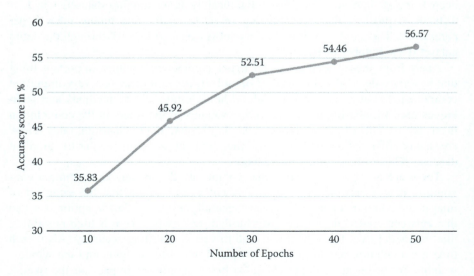

FIGURE 2.7 White blood cell classification accuracy chart for Case # 1.

Blood Cell Classification

From Figure 2.7, it can be observed that with the increase in the number of epochs lead to an increase in the accuracy score as well. But, from the 40th epoch, there is not a steep increase in the accuracy score. The average increase in the accuracy score from the 10 epochs to 50 epochs is 11.58%, with the highest classification accuracy reaching 56.57% for 50 epochs.

2.3.3.2 Case # 2

The architecture and the hyper parameters for the second case were the same as Case # 1, with the only difference being the optimizer. In the second case, instead of RMSProp, the Adam optimizer was used. The results obtained have been shown Figure 2.8.

From Figure 2.7 and Figure 2.8, it can be observed that there is a marginal increase in the white blood cell accuracy score when Adam optimizer was employed. The average increase in the accuracy score for Adam optimizer is also higher at 14.45%, with highest classification accuracy reaching 61.99% for 50 epochs. So, it can be said that the Adam optimizer is a more appropriately suited optimizer for the blood cell classification dataset under consideration.

2.3.3.3 Case # 3

In the third and final case for understanding the effectiveness of the convolution neural network in blood cell classification, a totally different and more complex architecture was designed.

First, a convolution layer was added with filter numbers as 64 with size of 3 × 3. Then, a max pooling layer was added with window size of 2 × 2. Then, again a pair of convolution layers and max pooling layers was added. The number of filters in this second convolution layer was 128 with size of 3 × 3. After the second set of

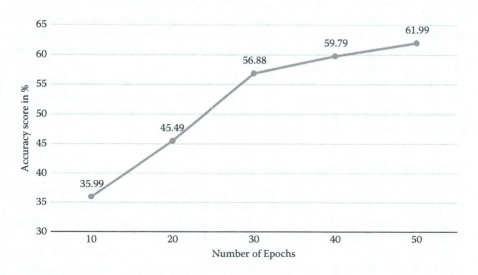

FIGURE 2.8 White blood cell classification accuracy chart for Case # 2.

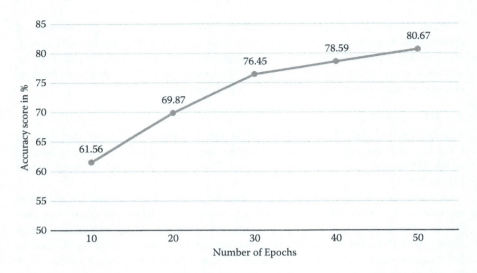

FIGURE 2.9 White blood cell classification accuracy chart for Case # 3.

convolution and max pooling layers, another set was added. The number of filters in the third convolution layer was 256 with the size of 3 × 3.

Three dense layers were added, with the first dense layer having 256 output neurons, the second dense layer having 128 output neurons, and the final dense layer having four output neurons. An Adam optimizer was used as well. The input size of the image was also increased to 256 × 256.

For the said architecture, the training and testing was carried out for five different epochs again. The results obtained have been showcased in Figure 2.9.

From Figure 2.9, it can be observed that the architecture taken in this case produces far better results than the previous simple architecture. For the proposed architecture, the white blood cell classification accuracy reaches a maximum of 80.67% for 50 epochs. This result has been produced without performing any type of pre-processing of the images.

From Figures 2.7, 2.8 and 2.9, two things can be observed. First, the Adam optimizer is far better suited for the dataset under consideration, and a more complex architecture, which gives the algorithm more legroom for better feature extraction, produces significantly better results.

2.4 CONCLUSIONS

The advances in the field of image processing and analysis have paved the way for these techniques to be employed in sensitive areas like the field of medical sciences. Medical imaging is playing a very crucial role in early diagnosis of diseases, which is in turn helping patients to receive early treatment.

One of the most important techniques that has seen great adaptation in many sectors is deep learning techniques. Deep learning techniques, like the recurrent neural network and convolution neural networks, have played a very positive role in many areas of business. One of the important areas where these deep learning techniques are being experimented with is the field of medical imaging.

In this chapter, an analysis of deep learning techniques, namely recurrent neural network and convolution neural network, has been carried out for the classification of white blood cells from the microscopic images. It has been observed that the convolution neural network is better suited for classification of white blood cells.

A detailed experimentation with different parameters and hyper parameters of convolution neural networks was carried out. From the results obtained from these experiments, it can be concluded that the choice of the optimizer plays a very important role in determining the accuracy of a deep learning model.

From the results obtained, it can also be concluded that if a larger legroom is provided to the convolution model by having a complex model with multiple convolution layers, then it can analyse the input images and extract features in a better way, thus resulting in better predictions. Fine-tuning of the hyper parameters like the number of filters, number of epochs, and the input image size is very crucial for obtaining better white blood cell prediction accuracy. So, it can be safely concluded that, for the said dataset, among the popular deep learning techniques, convolution neural network with a complex network of layers is best suited for the classification of white blood cells.

REFERENCES

[1] Mitchell, T. M. 1999. *Machine Learning*. McGraw-Hill Education.
[2] Tiwari, L., V. Awasthi, R. Miri, and G. R. Sinha. 2021. Detection of Lung Nodule and Cancer Using Novel Mask-3 FCM and TWEDLNN Algorithms. *Measurement*, Vol. 172, p. 108882. ISSN 0263-2241. 10.1016/j.measurement.2020.108882
[3] Nagarhalli, T. P., A. Save, and N. Shekokar. 2021. Fundamental Models in Machine Learning and Deep Learning. In *Design of Intelligent Applications Using Machine Learning and Deep Learning Techniques* ed. R. S. Mangrulkar, A. Michalas, N. Shekokar, M. Narvekar, P. V. Chavan, Chapman and Hall/CRC, pp. 13–36.
[4] Sarkale, A., K. Shah, A. Chaudhary, and T. P. Nagarhalli. 2018. An Innovative Machine Learning Approach for Object Detection and Recognition. *IEEE Second International Conference on Inventive Communication and Computational Technologies (ICICCT)*, pp. 1008–1010.
[5] Patterson, J. and A. Gibson. 2017. *Deep Learning: A Practitioner's Approach*. O'Reilly Publications.
[6] Chandrakar, R., R. Raja, and R. Miri. 2021. Animal Detection Based on Deep Convolutional Neural Networks with Genetic Segmentation. *Multimed Tools and Applications*, Vol. 5, pp. 1–14. 10.1007/s11042-021-11290-4
[7] Sahu, A. K., S. Sharma, and M. Tanveer. 2021. Internet of Things Attack Detection Using Hybrid Deep Learning Model. *Computer Communications*, Vol. 176, pp. 146–154, ISSN 0140-3664. 10.1016/j.comcom.2021.05.024
[8] Lundervold, A. S. and A. Lundervol. 2019. An Overview of Deep Learning in Medical Imaging Focusing on MRI. *Science Direct*, Vol. 29, Issue 2, pp. 102–127.
[9] Yamashita, R., M. Nishio, R. K. G. Do, and K. Togashi. 2018. Convolutional Neural Networks: An Overview and Application Iin Radiology. *Insights into Imaging*, Vol. 9, pp. 611–629, *Springer Open*.

[10] Eldridge, L. 2020. Types and Function of White Blood Cells (WBCs). https://www.verywellhealth.com/understanding-white-blood-cells-and-counts-2249217 (Accessed 31st March, 2021).
[11] Nall, R. 2020. What to Know about White Blood Cells. https://www.medicalnewstoday.com/articles/327446#summary (Accessed 31st March, 2021).
[12] Higuera, V. 2018. WBC (White Blood Cell) Count. https://www.healthline.com/health/wbc-count (Accessed 31st March, 2021).
[13] Territo, M. 2020. Overview of White Blood Cell Disorders. https://www.msdmanuals.com/en-in/home/blood-disorders/white-blood-cell-disorders/overview-of-white-blood-cell-disorders (Accessed 31st March, 2021).
[14] Blood Cell Images Dataset. https://www.kaggle.com/paultimothymooney/blood-cells (Accessed 31st March, 2021).
[15] Sherstinsky, A. 2020. Fundamentals of Recurrent Neural Network (RNN) and Long Short-Term Memory (LSTM) Network. *Elsevier Journal of Physica D: Nonlinear Phenomena, Special Issue on Machine Learning and Dynamical Systems, Vol. 404, Issue March 2020*, pp. 1–43.
[16] Patila, A. M., M. D. Patila, and G. K. Birajdar. 2020. White Blood Cells Image Classification Using Deep Learning with Canonical Correlation Analysis. *IRBM. Vol. 42, Issue 5*, pp. 378–389.
[17] Liang, G., H. Hong, W. Xie, and L. Zheng. 2019. Combining Convolutional Neural Network With Recursive Neural Network for Blood Cell Image Classification. *IEEE Access, Vol. 6*, pp. 36188–36197.
[18] Wibawa, M. S. 2018. A Comparison Study Between Deep Learning and Conventional Machine Learning on White Blood Cells Classification. *IEEE International Conference on Orange Technologies (ICOT)*, pp. 1–6.
[19] Throngnumchai, K., P. Lomvisai, C. Tantasirin, and P. Phasukkit. 2019. Classification of White blood Cell Using Deep Convolutional Neural Network. *IEEE 12th Biomedical Engineering International Conference (BMEiCON)*, pp. 1–4.

3 Evaluating the Effectiveness of the Convolution Neural Network in Detecting Brain Tumors

Tatwadarshi P. Nagarhalli and Sneha Mhatre
Vidyavardhini's College of Engineering and Technology, Vasai, Mumbai, India

Ashwini Save
Computer Engineering, VIVA Institute of Technology, Virar, Mumbai, India

Sanket Patil
Vidyavardhini's College of Engineering and Technology, Vasai, Mumbai, India

CONTENTS

3.1 Introduction...29
    3.1.1 Deep Learning and Convolution Neural Network 30
    3.1.2 Medical Imagery/Brain Tumor Detection 31
3.2 Related Work..31
3.3 Dataset 32
3.4 Evaluation of Convolution Neural Network Architectures..........................34
    3.4.1 Test Case # 1 34
    3.4.2 Test Case # 2 35
    3.4.3 Test Case # 3 37
3.5 Conclusions..38
References..39

3.1 INTRODUCTION

Recent times have seen very successful implementations of machine learning [1] and deep learning [2] techniques in many fields, including finance, security, automation, and medical imagery. Machine learning and deep learning are subfields of artificial intelligence that strive to learn new insights from the data examples rather than being given explicit instructions about what is to be accomplished. Especially, deep learning is being applied quite extensively in many domains and fields because of the deeper insights it is able to extract.

The deep learning techniques have been very successful in extracting deeper intuitions because of the liberty the techniques have with the extraction of relevant features. That is, feature extraction is done by the model itself in deep learning, whereas in machine learning the feature extraction is an important and independent task to be accomplished by the programmer [3].

The independence that the deep leaning techniques have in terms of extracting the features helps these techniques work in an environment where the input data given for training might vary extensively and feature extraction is difficult. For instance, if the model is to be developed from recognising and working with images, the extraction of features becomes difficult because different images will have different areas of interest for feature extraction. This is the major reason why, when it comes to working with complex data like texts, images, videos, and audio, deep learning techniques have been very popular and successful [4].

3.1.1 DEEP LEARNING AND CONVOLUTION NEURAL NETWORK

There are many machine learning and deep learning techniques that have been employed with great success in different domains of study and different areas for business. But, generally it has been observed that when it comes to working with the images, for better training and classification of the images, convolution neural networks have been more accurate compared to other deep learning techniques [5,6].

In the convolution neural network, the technique is able to extract deep intuition with the help of two layers, the convolution layer and the pooling layer [7]. Essentially, the feature extraction operation is handled by the convolution layer. This is accomplished by multiplying the weights of the convolution layer window with the image's pixel value and then averaging the results. Sliding this window across the entire image creates a new complicated or filtered image. A number of filtered images are generated for better feature extraction [8].

On the other hand, the pooling layer's goal is to minimize computational complexity. Following a feature extraction job with one convolution layer, a number of convoluted pictures are created, and a number of convolution layers can be applied. As a result, the number of filtered pictures grows at an exponential rate. Pooling, particularly the Max-pooling layer, is used to minimize computer complexity. Max-pooling lowers the size of filtered pictures while enabling just the most important characteristics to be processed further [9].

Generally, for the purpose of classification, the last layer added is a multi-layer perceptron network. This neural network can be a deep neural network as well. The depth of the neural network depends on the type of intuitions required.

3.1.2 Medical Imagery/Brain Tumor Detection

A brain tumor is an abnormal cell growth or mass in the brain [10]. Due to lifestyle changes and other factors, the number of cases of brain tumors has been increasing at a very exponential rate all over the world. Considering the case of the United States of America alone, the country has been witnessing a rapid growth in the number of brain tumor cases. According to one estimate, the number of cases per year has almost reached the figure of 100,000, and it has resulted in people spending billions of dollars [11].

Like any other disease, early detection and diagnosis of brain tumor plays a vital role in providing care for the patients and thus saving lives. Of the different ways in which the brain tumor can be detected is magnetic resonance imaging (MRI) or computed tomography (CT). MRI is a form of scan that produces detailed pictures of the inside of the body using high magnetic fields and radio waves [12]. On the other hand, CT scan creates a 3D picture of soft tissues and bones using a sequence of X-rays and a computer. The output of both these techniques ia in the form of a series of images [13].

The traditional method includes taking these scan images to the doctor specialising in neuroscience, who will study the images and make a proper diagnosis. This is a time-consuming process. For detecting a brain tumor, which can be identified easily, critical time is wasted in this process. Another factor here is the availability of a neurologist; there are very few of the doctors specialising in the said field.

So, in the quest for automating or reducing the human component in the process of brain tumor recognition, many systems and techniques have been proposed in recent times that act as an initial identifier before going to the specialist doctor. Here, it is important to note that the start-of-the-art technologies and systems act as a force multiplier and never a replacement. This is true for any artificial intelligence, machine learning or natural language processing system.

Hence, a system is required that can analyse these images and detect the presence or absence of brain tumors. This process of studying and analysing such medical images to produce an output is called medical imaging. Generally, in recent times the term *medical imaging* has become synonymous with the application of machine learning and deep learning on medical images to produce a prediction.

One of the recent studies and technologies that have proved to be very successful in the field of medical imaging has been deep learning techniques, especially convolution neural network. Convolution neural networks have been applied for brain tumor detection as well with varied success. And, considering the fact that convolution neural network has been designed to work well on images, the chapter proposes to evaluate the effectiveness of convolution neural network in detecting brain tumors.

3.2 RELATED WORK

Different methods have been proposed, and a combination of methods has been employed for the early and automatic detection of brain tumors using medical imagery. Earlier, the major focus on proper detection of brain tumors was using different image processing techniques like edge detection, image enhancement, and image super resolution, among others.

In recent times, the advent, research, and development of machine learning and deep learning has ensured that these technologies are employed in the field of medical imagery. This employment of machine learning and deep learning technologies has yielded mixed results. Because of this, the research and deployment of these learning techniques has been an ongoing process.

Papers such as [14] propose to perform brain tumor detection by using image processing and the machine learning technique. The paper proposes to make use of wavelet transform and the machine learning technique of support vector machines. The paper claims to have good results as well.

On the other hand, paper [15] proposes to make use of the conventional machine learning technique of the Hidden Markov Model or process along with the image processing technique of thresholding for the detection of brain tumors.

Similar to these two papers, many papers have either used both learning techniques and image processing techniques in combination or have used only deep learning techniques for accurate detection of brain tumors from medical images [16].

As it has been observed that deep learning techniques like the convolution neural network have been very successful in working with images for the purpose of image classification, it is imperative that this technique be evaluated for the detection of critical diseases like brain tumors, which are one of the classical examples of medical imagery.

3.3 DATASET

There are two very popular datasets that have been generally used for brain tumor detection. Both these datasets contain MRI images of the brain. The first dataset [17] contains a total of 253 MRI images, of which 155 images are brain tumor positive images, and the remaining 98 images do not have any traces of brain tumor. On the other hand, the second dataset [18] contains a total of 3000 MRI images, with 1500 images being positive for brain tumor, and the remaining 1500 are brain tumor negative images.

It was observed that many papers were working with the first dataset, with a total of 253 images. But, this is a very low number for a deep learning algorithm. Here, it has to be remembered that deep learning techniques require a large number of images for proper training. So, as the main technique under evaluation is the convolution neural network, it made sense to go for the second dataset with 3000 brain MRI images [19].

Generally, for training and testing the model, the same dataset is retained. That is, generally, out of the 3000 brain MRI images taken, around 80% of the images are taken for training and the remaining for testing. This is a good way of testing the model's accuracy. But, in the real world, the model might face and come across brain images that are of different quality and may vary from the type of images on which the model was trained. So, considering this fact it, was decided to train the Convolution Neural Network Model on the complete set of 3000 brain

Brain Tumor Detection

MRI images from the second dataset, and this trained model was tested on the brain MRI images from the first dataset containing 253 images.

It is believed that this way of training and testing where the training data and the testing data have been sourced from different sources would result in a more accurate representation of abilities and capabilities of the designed and developed model.

As a sample, normal brain MRI images and images with tumors have been shown. Figure 3.1 shows the sample MRI images of a normal brain that does not have any tumor, and Figure 3.2 showcases examples of a brain tumor detected in the MRI images.

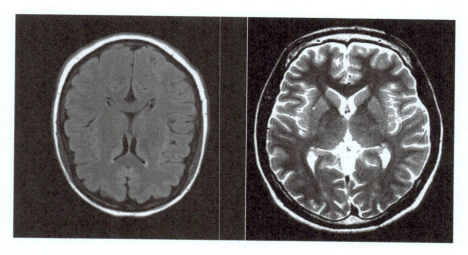

FIGURE 3.1 Normal brain MRI image.

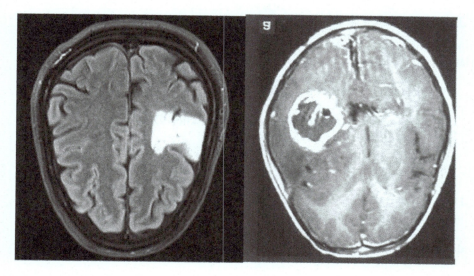

FIGURE 3.2 Brain images with tumor.

3.4 EVALUATION OF CONVOLUTION NEURAL NETWORK ARCHITECTURES

As mentioned, the amount of convolution layers, pooling layers, and dense layers can vary and be adjusted depending on the situation; this is a component of the study. Aside from that, the placement of these layers is a subject of investigation. For example, we may have a network design in which all of the convolution layers are utilized first, followed by the pooling layers, or we can have a network architecture in which everything is jumbled up. These are the many criteria that must be determined.

Aside from the aforementioned factors, there are a slew more hyper parameters to consider. For example, the input picture size, the filter size, the filter stride, the kind of activation function, and whether we want the filter size to be the same or lower are all major parameters that must be considered.

Similarly, when it comes to the pooling layer, it must first be determined which of the maximum, minimum, and average pooling layers are necessary. After that has been decided, the following step is to determine the window size and stride. The number of output neurons must be determined, even for the dense network.

The number of epochs is another hyper parameter that must be determined. Because deep learning techniques need a large number of training datasets, which might be difficult to come by, the same training set is used to fine-tune other hyper parameters. The number of epochs is the number of times the model is anticipated to run over the whole dataset.

Understanding these facts, three test cases have been derived with a different number of convolution layers, and maxpooling layers with different hyper parameter values. Each of these test cases have again been trained and tested with five different epochs.

3.4.1 TEST CASE # 1

As the first case, a simple convolution neural network has been designed. The architecture for the first test case follows in a sequential manner,

1. Convolution layer with 32 filters and with filter size of 3×3.
2. Maxpooling layer with window size 2×2.
3. Convolution layer with 64 filters and with filter size of 3×3.
4. Maxpooling layer with window size 2×2.
5. Fully connected layer or dense layer with 256 output neurons.
6. Dense layer with two output neurons.

Here, it has to be noted that the last fully connected layer has two output neurons, indicating the two classes of yes and no.

With this particular architecture, five different models were created with five different epochs (10, 20, 30, 40, and 50). Table 3.1 shows the accuracy results obtained at different epochs.

Brain Tumor Detection

TABLE 3.1
Accuracy results for Test Case # 1

Sl. No.	Number of Epochs	Validation Accuracy in %	Test Accuracy in %
1.	10	84.54	82.21
2.	20	84.23	87.35
3.	30	86.94	88.79
4.	40	87.97	89.87
5.	50	88.89	91.12

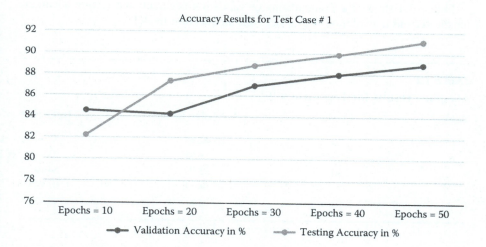

FIGURE 3.3 Validation and testing accuracy for Test Case # 1.

Figure 3.3 shows the comparisons of validation accuracy and testing accuracy for the present convolution neural network architecture model.

From Table 3.1 and Figure 3.3 it can be observed that the given convolution neural network model consistently produces good test accuracy results in excess of 80%. Also, it can be seen that with the increase in the number of epochs the accuracy results produced also increases.

3.4.2 Test Case # 2

As a second case, the complexity of the convolution neural network has been increased. The architecture for the second test case follows in a sequential manner,

1. Convolution layer with 64 filters and with filter size of 3×3.
2. Maxpooling layer with window size 2×2.

3. Convolution layer with 128 filters and with filter size of 2 × 2.
4. Maxpooling layer with window size 2 × 2.
5. Convolution layer with 128 filters and with filter size of 2 × 2.
6. Maxpooling layer with window size 2 × 2.
7. Convolution layer with 256 filters and with filter size of 2 × 2.
8. Maxpooling layer with window size 2 × 2.
9. Fully connected layer or dense layer with 256 output neurons.
10. Dense layer with two output neurons.

Similar to the first test case, even in the second case five different models were created with five different epochs (10, 20, 30, 40, and 50). Table 3.2 shows the accuracy results obtained at different epochs.

Figure 3.4 shows the comparisons of validation accuracy and testing accuracy for the second convolution neural network architecture model.

TABLE 3.2
Accuracy results for Test Case # 2

Sl. No.	Epochs	Validation Result	Test Result
1.	10	82.48	83.39
2.	20	86.32	83.94
3.	30	95.58	84.19
4.	40	96.13	87.72
5.	50	96.5	90.12

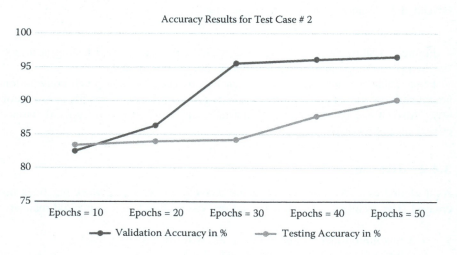

FIGURE 3.4 Validation and testing accuracy for Test Case # 2.

Brain Tumor Detection

From Table 3.2 and Figure 3.4 it can be observed that the given convolution neural network model consistently produces good test accuracy results in excess of 83% and reaches 90% for 50 epochs. Also, it can be seen that with the increase in the number of epochs, the accuracy results produced also increases.

Here, it is very important to note that there is a drastic difference in the validation and testing accuracies. There is a huge gap between the validation accuracy and testing accuracy. Although the validation accuracy clocks an accuracy result of more than 95%, the testing accuracy barely reaches the 90% mark for 50 epochs.

The drastic increase and major difference in the validation accuracy and testing accuracy also shows that with the increase in the number of epochs, the model is overfitting. When the validation accuracy is very high and the testing accuracy is comparatively on a lower side, it can be said that the learning model is overfitting. That is, the model is producing good results on the data it is trained on but is not able to perform classification on the unknown data.

3.4.3 Test Case # 3

As the final test case, the complexity of the convolution neural network has been increased further. The architecture for the third and final test case follows in a sequential manner,

1. Convolution layer with 64 filters and with filter size of 3×3.
2. Maxpooling layer with window size 2×2.
3. Convolution layer with 128 filters and with filter size of 2×2.
4. Maxpooling layer with window size 2×2.
5. Convolution layer with 128 filters and with filter size of 2×2.
6. Maxpooling layer with window size 2×2.
7. Convolution layer with 256 filters and with filter size of 2×2.
8. Maxpooling layer with window size 2×2.
9. Convolution layer with 256 filters and with filter size of 2×2.
10. Maxpooling layer with window size 2×2.
11. Fully connected layer or dense layer with 256 output neurons.
12. Dense layer with two output neurons.

Similar to the first two test cases, even in the third case five different models were created with five different epochs (10, 20, 30, 40, and 50). Table 3.3 shows the accuracy results obtained at different epochs.

Figure 3.5 shows the comparisons of validation accuracy and testing accuracy for the third convolution neural network architecture model.

From Table 3.3 and Figure 3.5, it can be observed that the given convolution neural network model consistently produces good test accuracy results in excess of 83% and reaching 92% for 50 epochs. Also, it can be seen that with the increase in the number of epochs, the accuracy of results produced also increases.

But, similar to the results seen in the second case, even here it can be seen that there is a vast gap between the validation accuracy and the testing accuracy, and

TABLE 3.3
Accuracy results for Test Case # 3

Sl. No.	Epochs	Validation Result	Test Result
1.	10	84.86	82.61
2.	20	87.28	85.33
3.	30	92.13	89.52
4.	40	94.79	91.89
5.	50	96.21	92.15

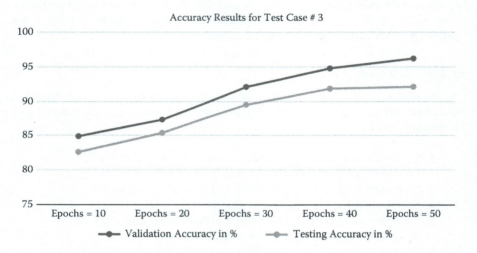

FIGURE 3.5 Validation and testing accuracy for Test Case # 3.

with the increase in the number of epochs, the gap between the validation accuracy and the testing accuracy widens further, with the validation accuracy being higher than the testing accuracy. This means that the model is again overfitting over the given training instances.

From the three test cases elaborated here, it can be seen that the convolution neural network produces very good accuracy results in detecting brain tumors. It can also be seen that with the increase in the complexity of the convolution neural network and with the increase in the number of epochs, the accuracy of the model also increases. So, it can be said that the convolution neural network is quite effective in detecting brain tumors from MRI images.

3.5 CONCLUSIONS

The advancements in image processing and analysis have cleared the door for these approaches to be used in sensitive fields such as medical science. Medical imaging plays a critical role in the early detection of illnesses, which aids patients in obtaining timely treatment. This is especially true with regards to the detection and diagnosis of brain tumors because of its fatality rate.

Deep learning techniques are one of the most essential approaches that have gained widespread adoption in a variety of fields. Deep learning techniques, such as recurrent neural networks and convolution neural networks, have shown to be quite useful in a variety of industries. Medical imaging is one of the most significant areas where deep learning techniques are being tested.

It has been observed that of the different deep learning techniques, convolution neural network is very highly regarded for processing of images. So, the chapter performs a detailed evaluation of the effectiveness of convolution neural network in detecting brain tumors from MRI images.

A total of 15 different models were created with different configurations for the evaluation. These 15 models were created with three different convolution neural network architectures and complexities. Also, in order to check the robustness of the convolution neural network, the training data and the testing data were sourced from different places.

It was observed that all the models were producing good testing accuracy results, ranging from 80% to 92%. But, it was observed that with the increase in the complexity of the convolution neural network and increase in the number of epochs, the gap between validation accuracy and testing accuracy also increased, with validation accuracy being on the higher side. This means that the model was overfitting.

This also means that if the testing was conducted from the same data source, then the accuracy of the results would have been on the higher side. And, as a different data source was taken for testing, this overfitting of the model came to light. So, measures should be taken to reduce the overfitting of the model. One measure that can be suggested to reduce the overfitting of the model is to increase the variety or impurity of the training dataset.

So, it can be concluded that the convolution neural network is a very effective technique for detecting brain tumors from MRI images. Also, more training data should be given in order to make the system more reliable and to reduce overfitting of the models.

REFERENCES

[1] Mitchell, T. M. 1999. *Machine Learning*. McGraw-Hill Education.
[2] Patterson, J. and A. Gibson. 2017. *Deep Learning: A Practitioner's Approach*. O'Reilly Publications.
[3] Trask, A. W. 2019. *Grokking Deep Learning*. Manning Publications.
[4] Sarkale, A., K. Shah, A. Chaudhary, and T. P. Nagarhalli. 2018. An Innovative Machine Learning Approach for Object Detection and Recognition. *IEEE Second International Conference on Inventive Communication and Computational Technologies (ICICCT)*, pp. 1008–1010.
[5] Chandrakar, R., R. Raja, R. Miri, U. Sinha, A. K. Kushwaha, and H. Raja. 2021. Enhanced the Moving Object Detection and Object Tracking for Traffic Surveillance Using RBF-FDLNN and CBF Algorithm. *Expert Systems with Applications*. 116306. ISSN: 0957-4174, 10.1016/j.eswa.2021.116306

[6] Padmanabhan, S. 2016. Convolutional Neural Networks for Image Classification and Captioning. https://web.stanford.edu/class/cs231a/prev_projects_2016/example_paper.pdf (Accessed 8th September, 2021).

[7] Nagarhalli, T. P., A. Save, and N. Shekokar. 2021. Fundamental Models in Machine Learning and Deep Learning. In *Design of Intelligent Applications Using Machine Learning and Deep Learning Techniques* ed. R. S. Mangrulkar, A. Michalas, N. Shekokar, M. Narvekar, P. V. Chavan, Chapman and Hall/CRC, pp. 13–36.

[8] O'Shea, K. T. and R. Nash. 2015. An Introduction to Convolutional Neural Networks. https://www.researchgate.net/publication/285164623_An_Introduction_to_Convolutional_Neural_Networks. (Accessed 29th August, 2021).

[9] Wang, B., Y. Liu, W. Xiao, Z. Xiong and M. Zhang. 2013. Positive and Negative Max Pooling for Image Classification. *IEEE International Conference on Consumer Electronics (ICCE)*, pp. 278–279.

[10] Editors. 2020. Brain Tumor. https://www.mayoclinic.org/diseases-conditions/brain-tumor/symptoms-causes/syc-20350084 (Accessed 8th September, 2021).

[11] Editors. 2020. The Potential: Changing the Face Of CNS-Disease Treatment. https://neonctech.com/potential/ (Accessed 8th September, 2021).

[12] Editors. 2020. Overview: MRI Scan. https://www.nhs.uk/conditions/mri-scan/ (Accessed 8th September, 2021).

[13] Editors. 2020. CT (Computed Tomography) Scan. https://my.clevelandclinic.org/health/diagnostics/4808-ct-computed-tomography-scan#:~:text=Computed%20tomography%20(CT)%20scan%20is,healthcare%20provider%20to%20diagnose%20conditions (Accessed 8th September, 2021).

[14] Gurbină, M., M. Lascu, and D. Lascu. 2019. Tumor Detection and Classification of MRI Brain Image using Different Wavelet Transforms and Support Vector Machines. *IEEE 42nd International Conference on Telecommunications and Signal Processing (TSP)*, pp. 505–508.

[15] Abdulbaqi, H. S., M. Zubir Mat, A. F. Omar, I. S. Bin Mustafa and L. K. Abood. 2014. Detecting Brain Tumor in Magnetic Resonance Images Using Hidden Markov Random Fields and Threshold Techniques. *IEEE Student Conference on Research and Development*, pp. 1–5.

[16] Handore, S. and D. Kokare. 2015. Performance Analysis of Various Methods of Tumor Detection. *International Conference on Pervasive Computing (ICPC)*, pp. 1–4.

[17] Brain MRI Images for Brain Tumor Detection. https://www.kaggle.com/navoneel/brain-mri-images-for-brain-tumor-detection (Accessed 8th September, 2021).

[18] Br35H: Brain Tumor Detection 2020: Brain Tumor Detection. https://www.kaggle.com/ahmedhamada0/brain-tumor-detection (Accessed 8th September, 2021).

[19] Raja, R., S. Kumar, S. Choudhary, and H. Dalmia. 2020. An Effective Contour Detection Based Image Retrieval Using Multi-Fusion Method and Neural Network, Submitted to Wireless Personal Communication, PREPRINT (Version 2) available at Research Square 10.21203/rs.3.rs-458104/v1

4 Implementation of Machine Learning in Color Perception and Psychology: A Review

Anusruti Mitra and Dipannita Basu
Department of Information Technology, Maulana Abul Kalam Azad University of Technology, West Bengal, India

Ahona Ghosh
Department of Computer Science and Engineering, Maulana Abul Kalam Azad University of Technology, West Bengal, India

CONTENTS

4.1 Introduction 42
    4.1.1 Motivation 42
    4.1.2 Related Works 42
    4.1.3 Contribution 43
4.2 Application Areas 44
    4.2.1 Food and Breed Hunting for Animals 44
    4.2.2 Application in Color Constancy 44
    4.2.3 Color Blindness Detection 44
    4.2.4 Sentiment Analysis Based on Color Attributes 45
    4.2.5 Application in Agriculture Using Color Classification 45
4.3 Deep Learning Methods Used in Color Psychology Analysis 45
    4.3.1 Conditional GAN 45
    4.3.2 Convolution Neural Network 45
    4.3.3 Bidirectional Long Short Term Memory 47
    4.3.4 Probabilistic Neural Network 48
    4.3.5 VGG-16 49
    4.3.6 DenseNet 50
4.4 Conclusion 50
References 50

DOI: 10.1201/9781003217091-4

4.1 INTRODUCTION

Color perception requires two elements, an illuminant and an observer, to grasp the object's color. Connivance between an illuminant and an observer is required, where the observer is the human encephalon [1]. The encephalon amasses the knowledge from each group of receptors, which gives rise to distinct perceptions of various wavelengths of light. In the human eye, cones and rods are not uniformly dispersed. The various wavelengths of visible light, such as red, are perceptible primarily at around 700 nanometers. The color violet's wavelength is 380 nanometers, which is short and frequent. Various colors invoke the feeling of happiness in the observer [2]. A happy color produces more satisfaction than perceived, an effect known as affective contrast enhancement. When a person views a dull-colored object, the light reflected activates the ocular process in the eye because contrasting illuminants have divergent spectral energy scattering, which is called color constancy—using light as a parameter of high- and low-intensity categorization of various colors carried out using machine learning and computer vision algorithms. Color blindness is a term given to the condition where individuals have trouble discriminating variations of color. It is not the shortcoming of the eye; instead, it is an obstruction of the brain, and the term is ambiguous because a person with color blindness is not blind. The association between color and emotion has possible comparability across cultures based on emotion surveys.

4.1.1 MOTIVATION

In today's age, there are also several objects that do not satisfy the color constancy theorem. Color has a deep relation with moods and emotion, and thus lifestyle. Whereas one area of red symbolizes love, another region captures hatred. Analyzing these emotions, a trend of color can be found. While reviewing other works, we felt that, to understand psychological needs, more theoretical analysis is required. Defining and controlling color scientifically is our main aim. Categorization of color takes communication in another place, which also motivates us. There are a lot of applications of color perception in human and animal life in various fields like agriculture and industrialization. Evaluating this application and creating a better path for future researchers is our main impetus.

4.1.2 RELATED WORKS

Freiders et al. organized a study to regulate the out-turn of color in a person's intellectual response. The study presented some pictures of participants in different colors and measured the reaction with the response of grayscale images. He used EEG and oximeter to determine the participant's heart rate, neural activity, and much more [3]. Ortis et al. illustrated the principle of general visual sentiment analysis systems through three points of view: emotional models, dataset definition, and feature design. The paper examines the structured formalization of the difficulty used for text assessment and talks about the appropriateness of visual sentiment analysis [4]. Watanabe et al. affirmed that communicative speech metrics are

Machine Learning in Color Perception

codified using its impersonation as a descriptor to evaluate its variation [5]. Patel et al. experimented and analyzed the color vision deficiency using machine learning algorithms and achieved some attributes related to other methods [6].

Gupta et al. examined the study in two parts. At first, human responses were taken to choose the subject using color perception methods to understand the mood associated with that color. Later, the study partly used a neural network to show the similarity between deep neural network models [7]. Ikeda et al. evaluated the study using a Stroop-like task, which was used to recognize color without facial expression to prove the influence of emotion with the help of recognizing the color [8]. Yoto et al. examined the frequency of human feelings using EEG that showed the alpha band; theta band has a higher frequency in red paper than blue paper, which indicates the state of anxiety represented in red-colored paper than blue-colored paper [9]. Rakshit et al. proposed the study based on four stimuli for color perception in which red-colored stimuli showed the maximum classification rate than yellow-colored stimuli [10].

Yong et al. explained the study to investigate facial expressions based on colors that will help others distinguish the purpose of a single color for every expression [11]. Rafegas et al. explored the study to show how color is encoded with the artificial network. The study is executed to determine the color selectivity index for each neuron to distinguish the activity of the neuron with color input response [12]. Olsson et al. described color constancy on chickens quantitatively by training the birds for color bigotry tasks and experimenting with them in changed radiance spectra to examine the highest illustration in which the color comes to constancy [13]. Forder et al. found evidence of color category. He analyzed the categorical relationship between colors and the variation in the frontal lobe [14].

Paulraj et al. used a protocol on visualization of colors like cyan, blue, magenta, black, red, green, white, and yellow to determine the high activity of the brain [15]. Jonauskaite et al. used a machine learning algorithm to assess the stability and attentiveness of color–emotion interrelation and the status of belonging to a country [16]. Bock et al. represented the first model with a precision that constructed spontaneous response; a new feature of mental imagery is produced on synesthetic perceptual experience [17]. Bloj et al. determined that color perception mostly occurs in 3-dimensional shape perception. The study experienced the light reflected from pale pink to deep magenta, incorporating the knowledge of mutual illumination with the physics of light [18]. Reis et al. reviewed a paper on the effect of color on object recognition. He used a meta-analysis to strongly encourage that color plays a role in object recognition. He recommended that the role of color should be taken into account for visual object identification [19].

4.1.3 Contribution

The contributions of our chapter can be summarized as follows

  i: Analyzing existing color perception algorithms is a novelty since no one has researched this domain before.
 ii: A possible road map for future researchers in the concerned domain.
iii: Detailed analysis of different applications in this domain.

The rest of the chapter is organized with the implementation perspectives, like the color perception used in the existing literature in the next section. The application areas and their prospects will be described in section 4.2. Different algorithms used in this domain have a detailed discussion in section 4.3. Finally, the concluding summary is mentioned in section 4.4, and a possible future direction has also been given.

4.2 APPLICATION AREAS

This section discusses the possible application areas of color perception analysis using machine learning tools.

4.2.1 Food and Breed Hunting for Animals

Colorful fruits and vegetables in a greenery hedge are very convenient to track down with dichromatic vision, and it is also effortless to find out whether a particular fruit is ripe or not from color perception [20]. Moreover, animals such as monkeys, ground squirrels, birds, insects, and many fish can catch sight of a good quality span of colors [21]. As an instance of this, bananas in a tree snatch the eyes of a monkey first rather than the eyes of a human [22].

4.2.2 Application in Color Constancy

Color constancy is a color that remains the same when light also reflects on another side. People have in-built color constancy in their minds [23]. A study conducted on chickens proved they also have color constancy. In the training phase, the chickens were put in a closed area with artificial white light and given access to food containers symbolized in three different colors: green, blue, and violet [24]. The chickens were also trained to have food by selecting the green-colored box. In the next phase of the study, when the white color was replaced by another artificial color such as red, the chickens also carried on with choosing the green-colored container. Using the same model on other animals, researchers for the first time found out color constancy capability in other animals. Color constancy assists birds in picking out equitable partners for life and breeding [25].

4.2.3 Color Blindness Detection

A class of situations that influences the perceptivity of color is termed *color blindness*. The deficit of color vision is utmost in red–green achromatic vision [26]. The Ishihara Plate test can detect color vision in schools or for medical purposes [27]. In a survey conducted on different types of color blindness in people, if a person is approved as red–green colorblind, it takes a prolonged time for them to perceive red and green. Still, the person can spot blues and yellows in no time, and the reverse happens with blue–yellow colorblindness [28]. From this study, we can say that both the colorblind people are confident in their comfort, and with this perception, they never claim false color [29].

4.2.4 SENTIMENT ANALYSIS BASED ON COLOR ATTRIBUTES

Perceptible content analysis has been dominant and challenging. Naturally, green as positive sentiment, red as negative sentiment, and gray as natural sentiment are treated [30]. Warm colors like red, yellow, and orange arrive at higher emotions, such as anger, love, and passion. Cool colors like purple, blue, and green are related to serenity, sadness, and indifference [31]. Colors can enhance intoxication states of mind and emotions. From these facts, we can conclude that sentiment analysis from any situation based on color attributes has been performed [32].

4.2.5 APPLICATION IN AGRICULTURE USING COLOR CLASSIFICATION

Color classification puts agriculture one step away from extending agriculture globally. Traditionally, color analysis is done manually using a colorimeter or a spectrophotometer [33]. The first phase of the process is individual and responsive, which causes exhaustion for the farmers [34]. The second stage is bound to estimate only a small-scale portion of the food, performing diligently to find a transparent vision of the shade of a food sample [35]. To better apply machine learning and artificial intelligence, we have restrained color analysis manually [36].

4.3 DEEP LEARNING METHODS USED IN COLOR PSYCHOLOGY ANALYSIS

The deep learning models applied in color psychology-based literature [37] are described in this section.

4.3.1 CONDITIONAL GAN

A generative adversarial network, abbreviated as GAN, is basically a deep learning-based generative model training architecture. A generator and a discriminator model are the main components of the architecture. The generator model creates new plausible examples that cannot be distinguished from the true examples of the dataset. And the discriminator component is in charge of determining if a particular image is real (taken from the dataset) or fraudulent (artificially generated).

The models are trained in a zero-sum or adversarial way so that increases in the discriminator come at the expense of the generator's competence and vice versa. GANs are good at picture fusion, which means they can generate new images given a target dataset. Some datasets contain extra information, like a class label, which should be utilized. Figure 4.1 has illustrated the working mechanism of conditional GAN architecture.

4.3.2 CONVOLUTION NEURAL NETWORK

A convolution neural network, abbreviated as CNN, is a deep learning architecture. CNN takes images as input and then analyzes a person's first perception and an amount of visual demonstration [38].

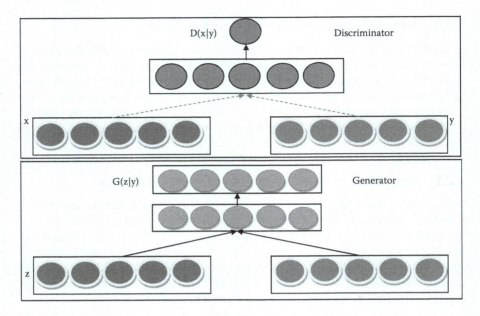

FIGURE 4.1 Pictorial illustration of conditional GAN architecture.

ALGORITHM: CNN

Step 1: Select an easy dataset on which it will help to clarify the image classification question. The dataset of emotions can bolster identifying different types of persons they are and thoughts of the mind.

Step 2: After choosing a dataset, we need to focus on training the data, which includes setting the correct path, designing labels, and changing the image size as required.

Step 3: Determine that the data contain pixel values and differentiate the index value from other images present in the dataset.

Step 4: The dataset should be mixed up so that the data can get trained properly.

Step 5: Allocating tab and its features using the neural metwork, classification methods will work to distinguish the shapes and patterns of the dataset.

Step 6: Reconcile X and transform tab to absolute data.

Step 7: Split X and Y coordinates to make use of CNN.

Step 8: Set, compile, and train the CNN replica.

Step 9: Use precision and scale representation [39].

Machine Learning in Color Perception

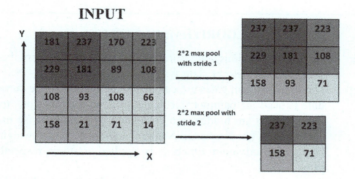

FIGURE 4.2 The working principle of CNN.

We can train the dataset using the CNN model from the above nine steps. The working mechanism of CNN is shown in Figure 4.2.

4.3.3 BIDIRECTIONAL LONG SHORT TERM MEMORY

The structure of the bidirectional long short term memory, abbreviated as Bi-LSTM, is straightforward. And Keras, a high-level API of Python considered for human beings, supports it. The wrapper takes a recurrent layer, the first layer of LSTM, as an argument. LSTM allows the merging of two different images [40]. Options are, first, the outputs are added together, and the outputs are multiplied. Then, the outputs can be added together using the Concat function to the next layer. Finally, the average of the outputs can be noted. The default mode is abstract. It is one type of sequence learning [41] shown in Figure 4.3.

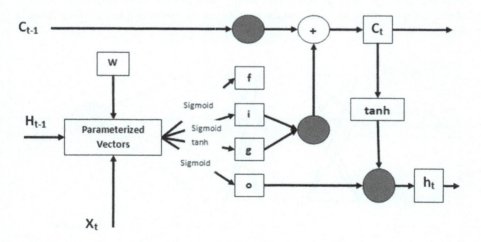

FIGURE 4.3 The working principle of Bi-LSTM.

> **ALGORITHM: BI-LSTM**
>
> Step 1: Firstly, the previous state is the hidden state, or the internal cell state, and the current state will be taken as input.
>
> Step 2: The four different gates are calculated differently as the current state is used for parametric calculation, and the other previous states, which are hidden, are calculated by element-wise multiplication with concerned vectors and their actual weights. Then, the activation function is applied to their respective gates.
>
> Step 3: Computing two states, the current internal state is evaluated by an element-wise multiplication vector of input gates and an input modulation gate. The previous internal state is enumerated again by using an element-wise multiplication vector.
>
> Step 4: To enumerate the current hidden state at first, we have to find the element-wise hyperbolic tangent of the present internal state and later operate element-wise multiplication with the output gate.

4.3.4 Probabilistic Neural Network

A probabilistic neural network, abbreviated as PNN, is a feedforward network. It is mainly applied to solve problems related to classification and pattern recognition. It has a three-layer architecture. It was first developed in 1990 by Spect. The four layers of PNN consist of one input layer, one pattern layer, one summation layer, and one decision layer. It is the probability of a sample belonging to a category. After defining PNN, we can sustain vectors on the network [42], as illustrated in Figure 4.4.

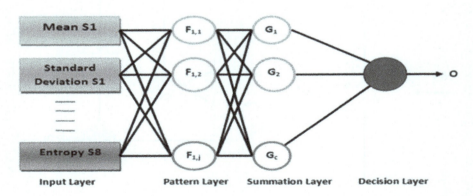

FIGURE 4.4 The working principle of PNN.

Machine Learning in Color Perception

ALGORITHM: PNN

Step 1: Interpret the input vector and sustain it to the Gaussian function for several categories.

Step 2: For every category of hidden nodes, calculate each Gaussian function.

Step 3: Next, cater for all Gaussian functions to a single output node for that category.

Step 4: In this stage, take the sum of all the inputs and multiply it with constant.

Step 5: Find out the maximum value of all outcoming nodes [43].

4.3.5 VGG-16

VGG-16 is an extension of convolution neural network and was used to win the Imagenet contest organized in 2014. This model has been considered to be the best architectural model for vision monitoring. The maximum pooling layer and disposition of convolution are the keys of the VGG-16 architectural model [44], as outlined in Figure 4.5.

ALGORITHM: VGG-16

Step 1: Import the necessary libraries required for the implementation of VGG-16.

Step 2: Train the model using an image generator concerning the exact size.

Step 3: After fitting, check the model summary.

Step 4: Using matplotlib, the sequence can be printed.

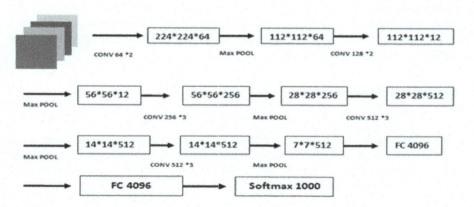

FIGURE 4.5 The working principle of VGG-16.

VGG-16 is very slow to train; in NVIDIA GPU, it took 2–3 weeks. Moreover, images trained on VGG-16 were 528 MB, which occupies storage and degrades the bandwidth [45].

4.3.6 DenseNet

To improve accuracy due to vanishing gradients and a deeper gap between the input and output layers, DenseNet came into existence in 2016 [46].

ALGORITHM: DENSENET

Step 1: Pre-activation is done with output feature maps of k-channels.
Step 2: Model complexity and size reduction are made by bottleneck layers.
Step 3: Global average pooling is calculated.
Step 4: Next, classification is done using a softmax classifier.
Step 5: Compression is done [47].

4.4 CONCLUSION

Color vision, a feature of visual perception, is the skill to perceive differences between light consisting of several wavelengths irrespective of light intensity. The researchers have made different attempts to identify and regulate the connectivity among the brain lobes during a person's color perception-based activities. Color psychology is the study of colors to determine human behavior. Color has an effect on experiences that aren't immediately noticeable, like the different flavors of food. Furthermore, colors include qualities that may cause people to experience specific sentiments. The present chapter reviewed and discussed different machine learning-based techniques and technologies applied in the literature involving color perception and psychology analysis, which can create a road map to future researchers in this domain so possible benefits and drawbacks can be explored further. A detailed description of the working mechanisms of different algorithms can be helpful in comparative study, and clinical applications like rehabilitation regarding color blindness will also benefit from it.

REFERENCES

[1] Özgen, E., 2004. Language, learning, and color perception. *Current Directions in Psychological Science*, *13*(3), pp. 95–98.
[2] Chowdhury, A., Dewan, D., Ghosh, L., Konar, A. and Nagar, A.K., 2020. Brain connectivity analysis in color perception problem using convergent cross mapping technique. Soft Computing for Problem Solving 2019. *Proceedings of SocProS 2019*, *1*(1138), p. 287.

[3] Freiders, S., Lee, S., Statz, D. and Kim, T., 2012. The influence of color on physiological response. *J. Adv. Stud. Sci*, *1*, pp. 1–12.

[4] Ortis, A., Farinella, G.M. and Battiato, S., 2020. Survey on visual sentiment analysis. *IET Image Processing*, *14*(8), pp. 1440–1456.

[5] Watanabe, K., Greenberg, Y. and Sagisaka, Y., 2014, December. Sentiment analysis of color attributes derived from vowel sound impression for multimodal expression. In *Signal and Information Processing Association Annual Summit and Conference (APSIPA), 2014 Asia-Pacific* (pp. 1–5). IEEE.

[6] Kumar, S., Jain, A., Shukla, A.P., Singh, S., Raja, R. and Rani, S., 2021. A comparative analysis of machine learning algorithms for detection of organic and nonorganic cotton diseases. *Mathematical Problems in Engineering*. Article ID 1790171, 18. 10.1155/2021/1790171

[7] Gupta, S. and Gupta, S.K., 2020. Investigating emotion-color association in deep neural networks. *arXiv preprint arXiv:2011.11058*.

[8] Ikeda, S., 2020. Influence of color on emotion recognition is not bidirectional: An investigation of the association between color and emotion using a stroop-like task. *Psychological Reports*, *123*(4), pp. 1226–1239.

[9] Yoto, A., Katsuura, T., Iwanaga, K. and Shimomura, Y., 2007. Effects of object color stimuli on human brain activities in perception and attention referred to EEG alpha band response. *Journal of Physiological Anthropology*, *26*(3), pp. 373–379.

[10] Rakshit, A. and Lahiri, R., 2016, July. Discriminating different color from EEG signals using interval-type 2 fuzzy space classifier (a neuro-marketing study on the effect of color to cognitive state). In *2016 IEEE 1st International Conference on Power Electronics, Intelligent Control and Energy Systems (ICPEICES)* (pp. 1–6). IEEE.

[11] Sinha, T.S., Patra, R.K. and Raja, R., 2011. A comprehensive analysis of human gait for abnormal foot recognition using neuro-genetic approach. *International Journal of Tomography and Statistics (IJTS)*, *16*(W11), pp. 56–73, ISSN: 2319-3339. http://ceser.res.in/ceserp/index.php/ijts

[12] Raja, R., Kumar, S., Md Rashid, 2020. Color object detection based image retrieval using ROI segmentation with multi-feature method. *Wireless Personal Communications*, *112*(1), pp. 169–192. Print ISSN: 0929-6212, online ISSN: 1572-834. 10.1007/s11277-019-07021-6

[13] Olsson, P., Wilby, D. and Kelber, A., 2016. Quantitative studies of animal colour constancy: Using the chicken as model. *Proceedings of the Royal Society B: Biological Sciences*, *283*(1830), p. 20160411.

[14] Forder, L., He, X. and Franklin, A., 2017. Colour categories are reflected in sensory stages of colour perception when stimulus issues are resolved. *PLoS One*, *12*(5), p. e0178097.

[15] Paulraj, M.P., Adom, A.H., Hema, C.R. and Purushothaman, D., 2012. Analysis of visual colour perception using EEG spectral features. *Karpagam Journal of Computer Science*, *6*(2), pp. 74–81.

[16] Jonauskaite, D., Wicker, J., Mohr, C., Dael, N., Havelka, J., Papadatou-Pastou, M., Zhang, M. and Oberfeld, D., 2019. A machine learning approach quantifies the specificity of color–emotion associations and cultural differences. *Royal Society Open Science*, *6*(9), p. 190741.

[17] Bock, J.R., 2018. A deep learning model of perception in color-letter synesthesia. *Big Data and Cognitive Computing*, *2*(1), p. 8.

[18] Bloj, M.G., Kersten, D. and Hurlbert, A.C., 1999. Perception of three-dimensional shape influences colour perception through mutual illumination. *Nature*, *402*(6764), pp. 877–879.

[19] Bramão, I., Reis, A., Petersson, K.M. and Faísca, L., 2011. The role of color information on object recognition: A review and meta-analysis. *Acta Psychologica*, *138*(1), pp. 244–253.

[20] Burtt Jr, E.H., 1981. The adaptiveness of animal colors. *BioScience*, *31*(10), pp. 723–729.

[21] Silvers, W.K., 2012. *The cOat Colors of Mice: A Model for Mammalian Gene Action and Interaction*. Springer Science & Business Media.

[22] Vorobyev, M. and Brandt, R., 1997. How do insect pollinators discriminate colors?. *Israel Journal of Plant Sciences*, *45*(2–3), pp. 103–113.

[23] Raja, R., Sinha, T. S. and Dubey, R.P., 2015. Recognition of human-face from side-view using progressive switching pattern and soft-computing technique. Association for the Advancement of Modelling and Simulation Techniques in Enterprises, *Advance B*, *58*(1), pp. 14–34, ISSN: 1240-4543.

[24] Patra R.K., Raja R. and Sinha T.S., 2018. Extraction of geometric and prosodic features from human-gait-speech data for behavioral pattern detection: Part II. In: Bhattacharyya S., Chaki N., Konar D., Chakraborty U., Singh C. (eds.), *Advanced Computational and Communication Paradigms. Advances in Intelligent Systems and Computing*. Vol. 706. Springer, Singapore, pp. 267–278. ISBN 978-981-10-8236-8.

[25] Renno, J.P., Makris, D., Ellis, T. and Jones, G.A., 2005, October. Application and evaluation of colour constancy in visual surveillance. In *2005 IEEE International Workshop on Visual Surveillance and Performance Evaluation of Tracking and Surveillance* (pp. 301–308). IEEE.

[26] Henmon, V.A., 1906. The Detection of Color-Blindness. *The Journal of Philosophy, Psychology and Scientific Methods*, *3*(13), pp. 341–344.

[27] Chandrakar, R., Raja, R., Miri, R., Patra, R.K. and Sinha, U., 2021. Computer succored vaticination of multi-object detection and histogram enhancement in low vision. *Int. J. of Biometrics. Special Issue: Investigation of Robustness in Image Enhancement and Preprocessing Techniques for Biometrics and Computer Vision Applications*, *34*, pp. 1–12.

[28] Van Staden, D., Mahomed, F.N., Govender, S., Lengisi, L., Singh, B. and Aboobaker, O., 2018. Comparing the validity of an online Ishihara colour vision test to the traditional Ishihara handbook in a South African university population. *African Vision and Eye Health*, *77*(1), pp. 1–4.

[29] Martin, C.E., Keller, J.O., Rogers, S.K. and Kabrinsky, M., 2000. Color blindness and a color human visual system model. *IEEE Transactions on Systems, Man, and Cybernetics-Part A: Systems and Humans*, *30*(4), pp. 494–500.

[30] Greenberg, Y., Tsuzaki, M., Kato, H. and Sagisaka, Y., 2010. Analysis of impression-prosody mapping in communicative speech consisting of multiple lexicons with different impressions. In *Proc. of Oriental International Committee for the Co-ordination and Standardisation of Speech Databases and Assessment Techniques, O-COCOSDA*.

[31] Nagata, N., Iwai, D., Wake, S.H. and Inokuchi, S., 2005, September. Non-verbal mapping between sound and color-mapping derived from colored hearing synesthetes and its applications. In *International Conference on Entertainment Computing* (pp. 401–412). Springer, Berlin, Heidelberg.

[32] Shoumy, N.J., Ang, L.M. and Rahaman, D.M., 2019. Multimodal big data affective analytics. In *Multimodal Analytics for Next-Generation Big Data Technologies and Applications* (pp. 45–71). Springer, Cham.

[33] Anami, B.S., Pujari, J.D. and Yakkundimath, R., 2011. Identification and classification of normal and affected agriculture/horticulture produce based on combined

color and texture feature extraction. *International Journal of Computer Applications in Engineering Sciences*, *1*(3), pp. 356–360.
[34] Pujari, J.D., Yakkundimath, R. and Byadgi, A.S., 2014. Recognition and classification of normal and affected agriculture produce using reduced color and texture features. *International Journal of Computer Applications*, *93*(11).
[35] Raja, R., Kumar, S, Choudhary, S. and Dalmia, H., 2021. An effective contour detection based image retrieval using multi-fusion method and neural network, Submitted to *Wireless Personal Communication,* PREPRINT Version-2 available at Research Square 10.21203/rs.3.rs-458104/v1
[36] Sandoval, J.R.M., Rosas, M.E.M., Sandoval, E.M., Velasco, M.M.M. and De Ávila, H.C., 2018. Color analysis and image processing applied in agriculture. *Colorimetry and Image Processing*, pp. 71–78.
[37] Chanda, K., Ghosh, A., Dey, S., Bose, R. and Roy, S., 2022. Smart self-immolation prediction techniques: An analytical study for predicting suicidal tendencies using machine learning algorithms. In Moh, Melody, Sharma, Kanta Prasad, Agrawal, Rashmi and Diaz, Vicente Garcia (Eds.) *Smart IoT for Research and Industry* (pp. 69–91). Springer, Cham.
[38] Bianco, S., Cusano, C., Napoletano, P. and Schettini, R., 2017. Improving CNN-based texture classification by color balancing. *Journal of Imaging*, *3*(3), p. 33.
[39] Bianco, S., Cusano, C. and Schettini, R., 2015. Color constancy using CNNs. In *Proceedings of the IEEE Conference on Computer Vision and Pattern Recognition Workshops* (pp. 81–89).
[40] Chavez, A.G., Mueller, C.A., Birk, A., Babic, A. and Miskovic, N., 2017, June. Stereo-vision based diver pose estimation using LSTM recurrent neural networks for AUV navigation guidance. In *OCEANS 2017-Aberdeen* (pp. 1–7). IEEE.
[41] Chandrakar, R., Raja, R., Miri, R., 2021. Animal detection based on deep convolutional neural networks with genetic segmentation. *Multimed Tools Application*, *10*, pp. 1–14. 10.1007/s11042-021-11290-4
[42] Ozcanli, A.K., Yaprakdal, F. and Baysal, M., 2020. Deep learning methods and applications for electrical power systems: A comprehensive review. *International Journal of Energy Research*, *44*(9), pp. 7136–7157.
[43] Raja, R., Nagwanshi, K.K., Kumar, S. and RamyaLaxmi, K., 2021. *Data Mining Technologies Using Machine Learning Algorithms*. Wiley & Scrivener. ISBN No: 13: 9781119791782.
[44] Raja, R. Sinha, T.S., and Dubey, R.P., 2018. Recognition of human-face from side-view using progressive switching pattern and soft-computing technique. Association for the Advancement of Modelling and Simulation Techniques in Enterprises.*Advance B*, *58*(1), pp. 14–34. ISSN: 1240-4543.
[45] Rangarajan, A.K. and Purushothaman, R., 2020. Disease classification in eggplant using pre-trained VGG16 and MSVM. *Scientific Reports*, *10*(1), pp. 1–11.
[46] Al-Bander, B., Williams, B.M., Al-Nuaimy, W., Al-Taee, M.A., Pratt, H. and Zheng, Y., 2018. Dense fully convolutional segmentation of the optic disc and cup in colour fundus for glaucoma diagnosis. *Symmetry*, *10*(4), p. 87.
[47] Tian, Y., Yang, G., Wang, Z., Li, E. and Liang, Z., 2019. Detection of apple lesions in orchards based on deep learning methods of cyclegan and yolov3-dense. *Journal of Sensors*, *2019*.

5 Early Recognition of Dynamic Sleeping Patterns Associated with Rapid Eyeball Movement Sleep Behavior Disorder of Apnea Patients Using Neural Network Techniques

Prateek Pratyasha and Saurabh Gupta
Department of Biomedical Engineering, National Institute of Technology, Raipur, Chhattisgarh, India

CONTENTS

- 5.1 Introduction..55
- 5.2 Methodologies 57
- 5.3 Data Collection...57
- 5.4 Preprocessing of Data and Feature Extraction Using Wavelet Packet Decomposition (WPD)..57
- 5.5 Feature Classification by Deep Neural Network (DNN) Classifier 61
- 5.6 Long Short Term Memory (LSTM) Technique...62
- 5.7 Results and Discussions 62
- 5.8 Conclusion..67
- References..68

5.1 INTRODUCTION

Sleep plays a role of temporary disconnection from vigilant activities and leads our mind and body towards quiescence. Sleeping behavior disorders are the cause of many neurological diseases, such as insomnia, sleep apnea, Parkinson's disease, etc.

DOI: 10.1201/9781003217091-5

Therefore, observation of sleeping behavior and sleeping stages is essential to diagnose sleep-related disorders [1]. The initial step is to classify the sleeping stage epochs from polysomnographic (PSG) datasets during sleeping hours. According to the time series epochs, the sleeping stages are classified into awaking stage, Rapid Non-Rapid Eye Movement (NERM) (sleeping stages: N1, N2, N3), and Eye Movement (REM) [2]. These behaviors are measured by using electrical stimulation on the brain scalp. The condition is predominantly associated with sleep apnea affecting spatial navigational memory, sympathetic, and cardiovascular activity; however, any sleeping pattern disorder within REM epochs may lead to degenerative neurological conditions that ultimately cause sleep paralysis, apnea, and cataplexy, which most commonly occurs in adults [3].

Rapid eye movement (REM) based sleep behavior disorder (SBD), or REM-SBD, is characterized by vocalizations and motor behaviors during the REM sleeping stage. Sleeping pattern classification and recognition has been a new challenge for researchers for decades. Recognition of sleeping patterns was started from direct measurement by EEG channels with transitional epochs of 20–30 seconds and a performance accuracy of 62.22% [3]. Later on, the classification efforts were involved with automated staging technologies, but they struggled to reach human-level efficiency. In [4], a high-order spectra (HOS) was used for classification of bi-spectrum and bi-coherence based extracted features, and the performance analysis curve was performed against the Gaussian mixture model (GMM) [5] with a performance accuracy of 88.7%. Later on, few have adapted frequency domain analysis after extracting the data by using the discrete energy separation algorithm (DESA) [6]. The features were classified on an iteration basis using amplitude envelope and instantaneous frequency (AM-FM) characteristics. Some of the statistical analysis included naive Bayes, K-nearest neighbor, decision tree, and random forest methods. Each method showed an average rate of performance accuracy of nearly 88.1%–90.5%.

Modifying the automated techniques, a number of machine learning algorithms have been studied to classify the complex non-linear PSG data by using fuzzy logic, dual tree network [7], hidden Markov models [7], support vector machines (SVM) [8], empirical mode decomposition [9] (EMD), and neural network techniques. The article [9] used a method from graph domain features and applied SVM to classify the wavelet-based features with a classification accuracy of 83%. However, neural networks pose human-level inter-agreement when tested against a small set of preselected PSG data. Few of the datasets were validated against real-time clinical data [10]. For a temporal sleeping pattern classification, neural network architecture segregated temporal context with 91% of sleeping pattern accuracy. In [11], a multi-channel EEG signal was used to classify the pattern by implicating convention neural network (CNN) architecture with finely-grained segments. However, it still has some bias of misclassification. To overcome this scenario, the author presented a modified artificial neural network model [12] that used more features in training data, which had feed-forward propagation. This improved the accuracy of classification but still was insufficient to store prolonged PSG data without any cross-validation. Hence, the recurrent neural network was used as a desired classifier. This makes the primary distinction of our objective successful.

In our proposed work, we applied an extended version of ANN, which is a deep neural network, to solve the data imbalance and temporal context classification problems during PSG data classification. Our objective was to classify the sleeping patterns of REM-SBD patients with maximum classification accuracy by carrying the minimum number of features. However, WPD has been proven as an evident feature extracted technique to decompose the raw data and provides minimum relevant features. Later on, it was fed to train the DNN model with multiple hidden layers. However, the convergence rate of the outcome was compared against a trained model with other existing LSTM networks.

5.2 METHODOLOGIES

In this work, we used open source PSG data from the MIT_BIH polysomnographic database to evaluate the proposed techniques. The data were down sampled to 50 Hz initially. The white noise from the data was filtered out by using a band-pass filter with 3-dB point at three different tones: first filtration at 50 Hz, the next one at 100 Hz, and the final one at 150 Hz. After the pre-processing phase, the data were decomposed using WPD. Wavelet energies, wavelet entropies, and statistical and average coefficients were calculated, and we found 14 features for each sampled data. Later on, the database was split into training and testing datasets. The classification model was framed on the training dataset by DNN, and finally the performance accuracy was validated on the testing dataset. The intact framework of the proposed methods is shown in Figure 5.1.

5.3 DATA COLLECTION

The database was collected from an open source known as MIT-BIH Polysomnographic Database [13,14]. Multiple physiologic signals, including EEG channels, were collected with respect to different sleep stages and disturbances during sleep apnea. The database contained polysomnographic data of 16 patients over 80 hours' worth of four-, six-, and seven-channel recordings. The entire process of training, testing, and validation was repeated 20 times for different orders of all the 16 subjects. The classification accuracy on target data was averaged to determine the regression curve. The diagnosis of RBD-SBD was confirmed by AASM criteria. All the subjects had a neurological observation with no signs of other neurological disorder other than sleep apnea. The subjects had no history of sleep complaints or nightmares. The recordings were carried out by using two EEGs, one EOG, and one EMG with all the referential EEG channels. The data are analyzed with a "no distraction" condition with a leave-on-out cross-validation. For each subject, the polysomnographic data are trained by applying DNN and LSTM, and are compared with its own target data. This was done in a sequential manner so that each network was initialized and trained on each of the subject data successively.

5.4 PREPROCESSING OF DATA AND FEATURE EXTRACTION USING WAVELET PACKET DECOMPOSITION (WPD)

Wavelet packet decomposition (WPD) is an extension of wavelet decomposition (WD) involving more than one basis and multiple bases. It improves the

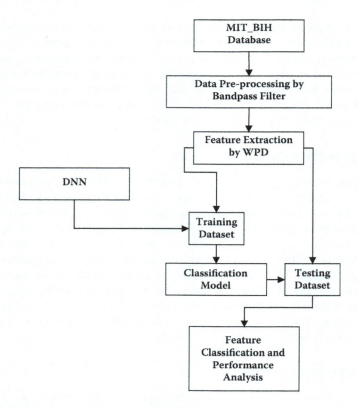

FIGURE 5.1 Framework of the proposed techniques.

classification performance and overcomes invariant time-frequency decomposition of discrete wavelet transform (DWT) [15]. The process of DWT starts with a common decomposition where the original signal is segmented into two coefficients, namely approximate coefficients (A) and detail coefficients (D) [16]. Both the classes are orthogonally complementary to each other, in which A carries the low-frequency dataset and D carries the high-frequency dataset. It allows some specific features to be decomposed in the time-frequency domain and transforms into a finite length wavelet [15]. However, decomposition is associated only with approximate coefficients, which are repeated while keeping the direct coefficients unchanged [16]. Nevertheless, the process of WPD decomposes the upper frequency bands of DWT in a defined frequency window. Figures 5.2(a) and (b) show the schematic diagram of DWT and WPD, respectively.

From the above figure, it is evident that WPD splits both the approximate and detail coefficients in a time-frequency domain. As compared to DWT, the methodology of WPD performs better frequency resolution of the signal while retaining the time-domain analysis at the top level. Therefore, for a decomposed signal, WPD extracts more features and makes it purposeful for further feature classification [17].

Early Recognition of Sleeping Patterns

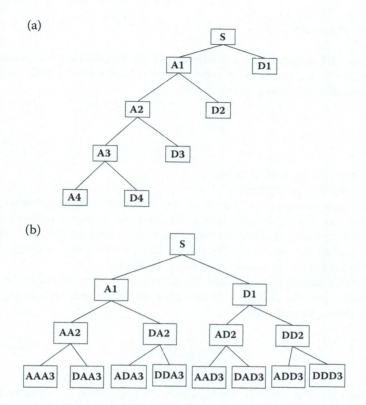

FIGURE 5.2 (a) Schematic diagram of DWT (b) Schematic diagram of WPD.

Let S be the signal of observance. When the signal recites at nth subspace at ith scale, it forms a reference wavelet packet as $S_{i,n}$. Here n is the frequency factor ($n = 0, 1, 2 \ldots i - 1$). Its corresponding orthonormal basis can be represented as $S_{i,k}^n(t)$ where k is the scale factor.

For n is even,

$$S_{i,k}^n(t) = \sum_k m_0(k) S_{i-1,k}^i \tag{5.1}$$

For n is odd,

$$S_{i,k}^n(t) = \sum_k m_1(k) S_{i-1,k}^i \tag{5.2}$$

Where $m_0(k)$ and $m_1(k)$ quadruple mirror filter as

$$m_1(k) = (-1)^{1-k} m_0(k) \tag{5.3}$$

Extracted Features:

1. **Wavelet Energy:** It is the summation of wavelet detailed coefficients in its squared form. The scale of wavelet energy varies over a wide range. The formulation of energy can be described as:

$$E(s) = \sum_{i=1}^{n} c_i d_i^2 \qquad (5.4)$$

Where c_i is the scaling factor.
d_i is the detailed coefficient.
The approximation coefficients of the first level of decomposition contain more energy that other levels of coefficients. As the signal is associated with high-frequency noise components along with the original data, it is preferable to consider the detail coefficients only. Here, we have obtained eight features from four levels of composition using WPD [18].

2. **Wavelet Entropy:** It indicated the quantitative energy stored in the decomposed signal. Several types of entropies are determined, such as log energy, threshold, Shannon etc.

$$E_n(s) = \sum_{i=1}^{n} S_i^2 \log(S_i^2) \qquad (5.5)$$

If entropy of a signal is observed over a norm p, then it is reformulated as:

$$E_n(s) = |S|^p \qquad (5.6)$$

Shannon entropy is formulated as:

$$E_n(s) = \sum_{i=1}^{n} S_i^2 \log(S_i^2) \qquad (5.7)$$

All the determined entropy values provide distinctive features of the signal and diminish the dimension of feature vectors for further processing [19].

3. **Statistical Decomposition Coefficients:** These feature vectors include average sub-bands, a set of zero crossing point numbers. These features possess simple computational time and have a lower feature dimensional space.

4. **Average Coefficients:** To extract the essential features, the sequence of centralized coefficients is reconfigured, and these coefficients are equal to one of the original discrete sequences of detailed coefficients. When the frequency of useful EEG signals is less than 50 Hz, sub-band means are determined on the basis of sampling frequency of the signal and centralized coefficients.

$$Mean_{SB} = \frac{2^n}{2^i} \sum_{j=0}^{50} S_i^n(j) \qquad (5.8)$$

Where 2^n is the sampling of each EEG channel. $j = (0, 1, \cdots \cdots 50)$ is the frequency range of initial features.

5.5 FEATURE CLASSIFICATION BY DEEP NEURAL NETWORK (DNN) CLASSIFIER

Deep neural networks (DNNs) are one of the most efficacious neural network tools to solve the complex classification tasks in various domains of artificial intelligence, computer vision, signal and image recognition, big data analytics, etc. A few decades ago, researchers introduced this technique in the domains of neuroscience, cognitive science, and brain–computer interface [20]. Before starting DNN, the foundation towards neural network has to be put at a glance. [21] A standard neural network comprises simple interconnected neurons and is expressed as a mathematical model. The model transforms either a single input neuron or a group of input neurons into a single output neuron via more than one intermediate neural channel. Each neuron is fed with weights to adjust the model during neural training to increase the performance efficacy of output. The resultant is compared against a pre-assumed value to determine the error. DNN has many neuron layers where the output of the current layer feeds input to the next layer [22]. This scheme enables a nonlinear and hierarchical relationship among the neurons. The layered framework of DNN is shown in Figure 5.3.

The architecture of DNN performs the process of computation in many folded layers. The input data samples $X = [x_1, x_2, \ldots x_n]$ for n prediction time. The input is

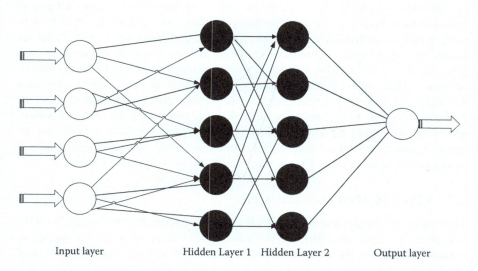

FIGURE 5.3 Layered framework of DNN.

linearly combined with some parameters like weights $\omega = [\omega_1, \omega_2 \ldots \omega_n]$ and a bias b_n. The pre-activation function is also based on the two mentioned parameters as:

$$a(x) = \omega(x) + b_n \tag{5.9}$$

The hidden layers are shown as $h(d)$. The layer-wise propagation method decomposes the signals into a single output layer as $y(x)$. The relationship of the output layer with input and the hidden layer is shown as:

$$y(x) = f\left[a^n(x)(h^n(a^{n-1}(\cdots \cdots \cdots (h^1(a^1(x))))))\right] \tag{5.10}$$

It is evident that model training depends on several parameters. However, both training and test datasets must be indepedently and identically distributed regardless of how the network predicts data classification. Nevertheless, validation of a trained network model is not conducted rigorously, but rather similar to that. It indicated how the data are generalized for testing without merging them with trained data. The drawback of this technique is that every notion of DNN is a black box. In the field of neuroscience, this drawback can't be ignored as classification performance analysis of neuro-physiological data demands the utmost importance [23].

5.6 LONG SHORT TERM MEMORY (LSTM) TECHNIQUE

Long short term memory (LSTM) is the extended temporal network model of the recurrent neural network (RNN) used in the deep learning method as a feedback connection. It overcomes the limitation of DNN by modeling the temporal context, having a temporal scope as conditional random fields, performing temporal inference over all the features, and emerging as a promising solution for sleeping pattern classification. A LSTM unit comprises a cell of memory piles to store long-term information in both the time and frequency domains; an input of recurrent time is used, which is multiplicative and protective from perturbation by other irrelevant inputs, an output based on current input values, and finally an internal memory gate [24]. The basics of LSTM have been described in other articles in detail. By taking multiple LSTM layers in parallel, data memorization in deeper temporal structures has increased. One LSTM unit is applied in forward direction and another one in backward direction to determine the sleep scoring label during the past and present epochs of each night [25]. This technique is evident to capture the desired features of sleeping patterns as well as increase the performance index of the short-term recurrent model independent of time. The architecture of LSTM is shown in Figure 5.4.

5.7 RESULTS AND DISCUSSIONS

The original data is down-sampled to 50 Hz, and a band-pass filter with range 9–14 Hz is applied to the signal to filter out the noise. A four-level db 2 WPD is applied to extract the features from the windowed signal. Each subject contains 5600 data samples. So, the dimension of the intact dataset is 179,200 samples (16 subjects × 2 EEG channels × 5600 samples). The dataset is normalized within the range of [0, 1].

Early Recognition of Sleeping Patterns

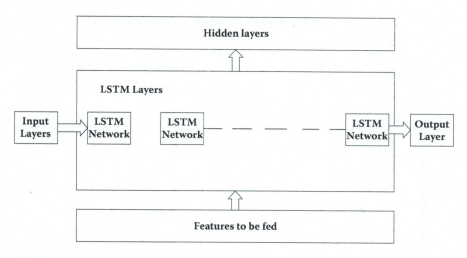

FIGURE 5.4 Architecture of LSTM network model.

Some of the spatial temporal features, such as wavelet energy, wavelet entropy, and statistical and average decomposition features, are extracted using WPD, out of which 896 features (14 features × 16 × 4 level decomposition) have been considered. The comparison between the original signal and the decomposed signal is depicted in the figure, where the red marks represent the original signal, whereas the black ones represent the noise-free decomposed signal. Figure 5.5 shows a comparative graph between the original and decomposed signals.

Now, the extracted dataset is separated into training and testing datasets; 80% of the samples (i.e., 142,640 samples) are chosen as the training dataset, and the rest of the 20% of data (i.e., 35,664 samples) are considered as the testing dataset. The network applied here consists of one input layer, two hidden layers, and one output layer. The network is trained using a standard by using a novel back-propagation method selecting five random training samples. The performance accuracy is achieved after terminating the training process of 1000 iterations, as graphed in Figure 5.6.

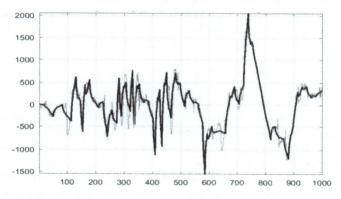

FIGURE 5.5 Original EEG signal and decomposed signal after applying WPD.

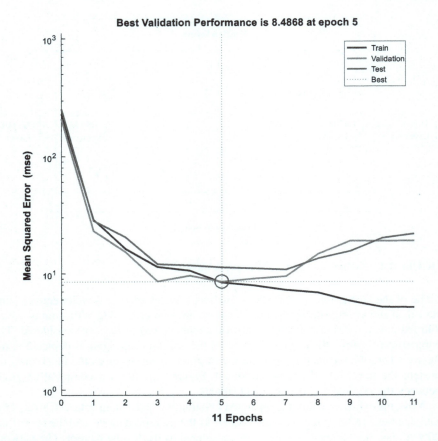

FIGURE 5.6 Performance validation curve.

According to the figure, mean square error (MSE) has been determined to be 8.4868 at epoch 5.

MSE represents the short- term effacacy of training model by taking the difference between observed values and predicted values. The lower the MSE, the more accurate the evaluation. The MSE is calculated as:

$$\text{MSE} = \frac{1}{n} \sum_{i=1}^{n} \left| y(x) - \hat{y}(x) \right|$$

An MSE closer to zero indicates that the training model is effectively fitted for classification. If we increase the number of fitted coefficients in the training and testing model, the value of R increases, though the curve fitting may not increase model performance accuracy.

The three plots in Figure 5.7 represent regression curve analysis of training data, validation data, and test data, respectively. The dotted line in each plot shows the ideal outcome that is result − output = target.

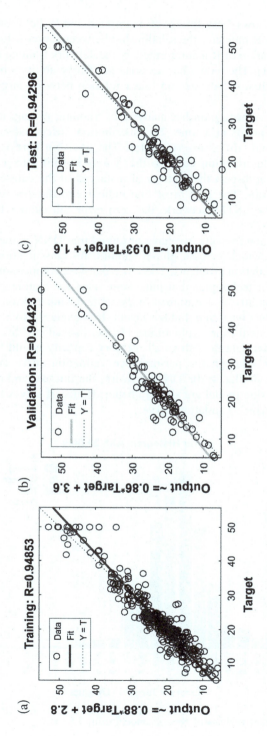

FIGURE 5.7 (a) Regression analysis of training data (b) Regression analysis of validation data (c) Regression analysis of testing data.

The solid line shows best fitting of regression curve between targets and output. The value of R indicates the relationship between target values and output value. The determination of R is interpreted by the network training model, which ranges between [0, 1]. The model has accurate predictive ability when R tends to 1. If R tends to 0, then it indicates no linear relation between target values and output value.

Figure 5.8 shows a histogram bar analysis of training, testing, and validation data by taking 300 points at a time. The distribution differs substantially in all the datasets and yields different error rates. The total bin here is divided into 20 intervals. The classification error in the X-axis is split into a bin range of 1–1.5. The Y-axis represents the number of instances of the dataset that lie in a particular range of bin. The zero error line indicates to the zero error value of the error axis. In the LSTM network, the zero error falls under the bin at the center point of 0.4674.

From all the obtained results, we have observed that DNN is the extension of conventional ANN, which gives a global solution for classification, regression, recognition, and prediction problems. There is a pairwise comparison of MSE and R values of different techniques that have been used for feature extraction and classification sequentially, as mentioned in the table. Compared to other classical neural networks, DNN has many hidden layers. It is expected to be faster to use the DNN model for real-time applications. Hence, instead of PSG data classification, a lower model dataset with small memory capacity would be efficient in implementing DNN. Though the performance validation curve and regression curve are relatively similar in the DNN network, the internal and external parameters in the regression model are slightly better. Hence, it solves the overfitting issue of the performance validation curve.

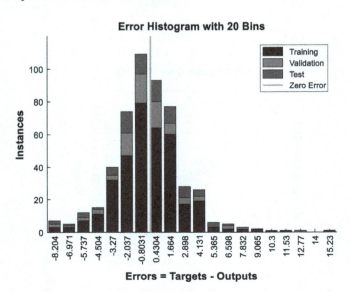

FIGURE 5.8 Histogram analysis of network model using LSTM.

TABLE 5.1
Comparison list of performance analysis of different feature extraction techniques followed by neural networks

Sl. No.	Different Neural Network Models	MSE		R Value	
		Training Data	Testing Data	Training Data	Testing Data
1	LDA+ANN	0.49682	0.46352	0.78761	0.80347
2	MLP+PNN	0.30452	0.64482	0.86423	0.88158
3	PCA+CNN	0.52358	0.46443	0.91753	0.89034
4	CWT+CNN	0.44235	0.46511	0.89315	0.86423
5	GDA+RNN	0.28621	0.51525	0.94712	0.91742
6	PCA+PNN	0.74516	0.74324	0.88217	0.86324
7	Fuzzy+NN	0.45714	0.64121	0.81932	0.90477
8	DT+KNN	0.42140	0.53354	0.85721	0.87421
9	WPD+LSTM (Proposed Method)	0.24527	0.13415	0.94415	0.93064
10	WPD+DNN (Proposed Method)	0.11485	0.24198	0.94853	0.94296

Table 5.1 shows some of the existing neural network models along with feature extraction techniques with a comparison of MSE and R values. From the table, it is clearly observed that DNN and LSTM models have slightly smaller MSEs and a specified R values. All the results reported are validated for training and testing datasets only. A neural networking training with a LSTM model shows an average MSE rate of 0.189925, and the corresponding regression value is 0.93607, whereas performance accuracy with DNN shows an average MSE rate of 0.178415 with corresponding R value of 0.945745. The comparison study signifies that DNN is outperformed when a smaller number of features are taken. However, considering a large number of training datasets, the LSTM model shows a better rate of convergence.

5.8 CONCLUSION

In this proposed work, a combined approach of wavelet packet decomposition and deep neural network is used for sleeping pattern recognition of REM-SBDr. Initially, some brief properties of neural network modeling methods and their related basic concepts are introduced. This DNN technique is an attempt to model the complex non-linear neural data, but this scheme can't be used for model selection. The trained model shows better performance accuracy compared to other neural network models. The obtained results are compared with other existing neural network techniques in terms of mean square error and coefficient of determination. The value of R is found to be near 0.94 for both training and testing data. From the obtained error histogram, it is evident that LSTM is valid only when we deal with a large number of features without any extraction techniques. However, the rate of convergence of a dataset trained with LSTM is outperformed with efficient predictive performance. Therefore, selection of the best neural network classifier is totally dependent on the problem objective and dimensionality of the dataset.

REFERENCES

[1] A. Sors, S. Bonnet, S. Mirek, L. Vercueil, and J.-F. Payen, "A convolutional neural network for sleep stage scoring from raw single-channel EEG," *Biomedical Signal Processing and Control*, vol. 42, pp. 107–114, 2018.

[2] R. Tripathy and U. R. Acharya, "Use of features from RR-time series and EEG signals for automated classification of sleep stages in deep neural network framework," *Biocybernetics and Biomedical Engineering*, vol. 38, pp. 890–902, 2018.

[3] K. A. Aboalayon, W. S. Almuhammadi, and M. Faezipour, "A comparison of different machine learning algorithms using single channel EEG signal for classifying human sleep stages," in *2015 Long Island Systems, Applications and Technology*, 2015, pp. 1–6.

[4] B. Das, M. Talukdar, R. Sarma, and S. M. Hazarika, "Multiple feature extraction of electroencephalograph signal for motor imagery classification through bispectral analysis," *Procedia Computer Science*, vol. 84, pp. 192–197, 2016.

[5] E. Thomas, A. Temko, G. Lightbody, W. Marnane, and G. Boylan, "Gaussian mixture models for classification of neonatal seizures using EEG," *Physiological Measurement*, vol. 31, p. 1047, 2010.

[6] R. R. Sharma, P. Varshney, R. B. Pachori, and S. K. Vishvakarma, "Automated system for epileptic EEG detection using iterative filtering," *IEEE Sensors Letters*, vol. 2, pp. 1–4, 2018.

[7] A. Singh and N. Kingsbury, "Efficient convolutional network learning using parametric log based dual-tree wavelet scatternet," in *Proceedings of the IEEE International Conference on Computer Vision Workshops*, 2017, pp. 1140–1147.

[8] J. Nalepa and M. Kawulok, "Selecting training sets for support vector machines: A review," *Artificial Intelligence Review*, vol. 52, pp. 857–900, 2019.

[9] E. Alickovic, J. Kevric, and A. Subasi, "Performance evaluation of empirical mode decomposition, discrete wavelet transform, and wavelet packed decomposition for automated epileptic seizure detection and prediction," *Biomedical Signal Processing and Control*, vol. 39, pp. 94–102, 2018.

[10] S. K. Satapathy, S. Sharathkumar, and D. Loganathan, "Automated sleep staging using convolution neural network based on single-channel EEG signal," In *Communication and Intelligent Systems*, Springer, Singapore, 2021, pp. 643–658.

[11] R. Miotto, F. Wang, S. Wang, X. Jiang, and J. T. Dudley, "Deep learning for healthcare: Review, opportunities and challenges," *Briefings in Bioinformatics*, vol. 19, pp. 1236–1246, 2018.

[12] D. J. Hemanth, "EEG signal based modified kohonen neural networks for classification of human mental emotions," *Journal of Artificial Intelligence and Systems*, vol. 2, pp. 1–13, 2020.

[13] R. Wei, X. Zhang, J. Wang, and X. Dang, "The research of sleep staging based on single-lead electrocardiogram and deep neural network," *Biomedical Engineering Letters*, vol. 8, pp. 87–93, 2018.

[14] V. Gurrala, P. Yarlagadda, and P. Koppireddi, "Detection of sleep apnea based on the analysis of sleep stages data using single channel EEG," *Traitement du Signal*, vol. 38, p. 2, 2021.

[15] W. Ting, Y. Guo-Zheng, Y. Bang-Hua, and S. Hong, "EEG feature extraction based on wavelet packet decomposition for brain computer interface," *Measurement*, vol. 41, pp. 618–625, 2008.

[16] V. V. Krishnan and P. B. Anto, "Features of wavelet packet decomposition and discrete wavelet transform for malayalam speech recognition," *International Journal of Recent Trends in Engineering*, vol. 1, p. 93, 2009.

[17] J. Kevric and A. Subasi, "Comparison of signal decomposition methods in classification of EEG signals for motor-imagery BCI system," *Biomedical Signal Processing and Control*, vol. 31, pp. 398–406, 2017.

[18] C. Uyulan, T. T. Ergüzel, and N. Tarhan, "Entropy-based feature extraction technique in conjunction with wavelet packet transform for multi-mental task classification," *Biomedical Engineering/Biomedizinische Technik*, vol. 64, pp. 529–542, 2019.

[19] E. Avci, I. Turkoglu, and M. Poyraz, "Intelligent target recognition based on wavelet packet neural network," *Expert Systems with Applications*, vol. 29, pp. 175–182, 2005.

[20] W. Samek, A. Binder, G. Montavon, S. Lapuschkin, and K.-R. Müller, "Evaluating the visualization of what a deep neural network has learned," *IEEE Transactions on Neural Networks and Learning Systems*, vol. 28, pp. 2660–2673, 2016.

[21] H. Liu, X. Mi, and Y. Li, "Smart deep learning based wind speed prediction model using wavelet packet decomposition, convolutional neural network and convolutional long short term memory network," *Energy Conversion and Management*, vol. 166, pp. 120–131, 2018.

[22] D. Bau, J.-Y. Zhu, H. Strobelt, A. Lapedriza, B. Zhou, and A. Torralba, "Understanding the role of individual units in a deep neural network," *Proceedings of the National Academy of Sciences*, vol. 117, pp. 30071–30078, 2020.

[23] A. Canziani, A. Paszke, and E. Culurciello, "An analysis of deep neural network models for practical applications," *arXiv preprint arXiv:1605.07678*, 2016.

[24] A. Sherstinsky, "Fundamentals of recurrent neural network (RNN) and long short-term memory (LSTM) network," *Physica D: Nonlinear Phenomena*, vol. 404, p. 132306, 2020.

[25] X. Song, Y. Liu, L. Xue, J. Wang, J. Zhang, J. Wang, et al., "Time-series well performance prediction based on Long Short-Term Memory (LSTM) neural network model," *Journal of Petroleum Science and Engineering*, vol. 186, p. 106682, 2020.

6 Smart Attendance cum Health Check-up Machine for Students/Villagers/Company Employees

Pranjal Patel, Shriram Sharma, Pritesh Sutrakar, Hemant Kumar, Devender Pal Singh, and Menka Yadav

Electronics and Communication Engineering, Malaviya National Institute of Technology, Jaipur, India

CONTENTS

- 6.1 Introduction 72
- 6.2 Overview of Various Possible Smart Attendance Systems 72
  - 6.2.1 Proposed Solution 73
  - 6.2.2 Face Recognition System 73
    - 6.2.2.1 Face Detection 74
    - 6.2.2.2 Face Recognition 74
- 6.3 Health Parameters and Stress Detection 75
  - 6.3.1 Eye Blink Detection 75
  - 6.3.2 Emotion Detection 76
- 6.4 Physiological Parameters and Sensors Description 76
  - 6.4.1 Body Temperature 76
  - 6.4.2 Blood SpO2 Level 77
  - 6.4.3 Heart Rate 77
- 6.5 Hardware and Sensors Used 77
  - 6.5.1 MLX 90614 Temperature Sensor 77
  - 6.5.2 MAX 30100 Heart Rate and Blood SpO2 Sensor 77
  - 6.5.3 Arduino UNO 78
- 6.6 Software Tools 79
  - 6.6.1 Arduino IDE 79
  - 6.6.2 Tkinter 79
- 6.7 Outcomes and Result 79
  - 6.7.1 Face Recognition 79

DOI: 10.1201/9781003217091-6

    6.7.2   Emotion Recognition and Eye-Blink Rate Detection 80
    6.7.3   Data from Sensors 81
6.8   Conclusion 82
Acknowledgement 82
References 82

6.1 INTRODUCTION

With the boom in digitalization and the virtual world of social media, people are generally insecure about their social acceptance. It makes the wrong impression on students that their friends are leading a more successful, comfortable, and happier life compared to theirs. More time on social media may also increase the risk to students. They can become an easy victim of cyberbullying or fraud, which can cause depression [1,2]. Also, many online games such as "The Blue Whale game" make students their target because they are emotionally weak or may have disturbed mental conditions. People generally seem to be ignorant of their medical condition. Sometimes they are not sure about the cause of their continuous sadness and sorrow, and as a result, they get in a state of mind where they get suicidal tendencies and fears [3]. In general, a student who does know that they are experiencing depression or having symptoms may be hesitant to receive help from anyone because of the "humiliation" that comes with depression [4,5]. It is best to identify the signs of depression at the initial stages. Facial expressions are a very good representation of the mood of a person without the need to communicate [6]. Many studies have been carried out and are ongoing for finding facial expressions that are related to depression [7–9]. Our current system will use various image processing techniques for face detection to mark student attendance, feature extraction, and classification of these facial features. The system will be trained with features of depression. Then, front-facing videos of students will be captured using a web camera system. We will then try to find some parameters that can help us in health monitoring and can act as markers for stress. It will tell us the stress level and other parameters like blink rate [10].

6.2 OVERVIEW OF VARIOUS POSSIBLE SMART ATTENDANCE SYSTEMS

The attendance of students going to schools is being maintained by schools and colleges as a record of their presence. The old, manual attendance marking procedure is very time consuming and hard to maintain and manage for every student. There is a need for a system/solution that can solve this issue of the manual attendance marking procedure. As we are in the modern digital world, biometrics technologies can be used since they affect people's daily life in various ways.

Biometric-based methods generally make use of features like fingerprints, face iris, retina patterns, palm prints, the voice of a person, and a handwritten signature for authentication of an individual [11,12]. These methods are considered to be very effective versus older security processes, such as a password or ID cards, because these methods involve physical data of a person. Biometric authentication uses data

Smart Attendance cum Health

taken from a person, barcode readers, or a radio frequency identification (RFID) system [13]. A biometric system consists of two stages: first, the enrollment phase, and second, the recognition phase [14]. Enrollment consists of recording the biometric trait of a person, storing the features in its database, as well as an identifier associated with the individual. The recognition process consists of taking the biometric feature of a person, extracting the identifier from it, and checking it in the database to find a possible match of record.

We have used a simple and effective solution in our machine: a face detection and recognition based system to mark attendance. Face recognition has various advantages over other biometric methods. Most biometric methods will need some action to be performed by the individual, whereas an attendance system based on face recognition will not require much involvement of the user because images of a person can be taken from a distance through a camera [15,16]. Overall, face recognition systems require minimal involvement and hence do not expose the user to any germs or viruses that may be present in a system that has multiple users. Our system is backed by a database. We implemented a design that also records some body parameters for health condition monitoring, and from the face, it detects stress level and emotion to integrate with the real-time operating student attendance system.

6.2.1 PROPOSED SOLUTION

Our machine will use face recognition for attendance, and it is integrated with the below-discussed features to provide an effective solution for health monitoring. The face recognition-based attendance system is integrated with:

- Eye blink rate detection
- Facial emotion detection and stress-level determination
- Arduino and sensor-based solution for health monitoring integrated with the attendance system:
  a. Temperature
  b. Heart beat
  c. Blood SpO2 level
  d. GSR sensor

6.2.2 FACE RECOGNITION SYSTEM

Every human face is unique in itself and can be used as one's identity. Therefore, face recognition is also considered to be a biometric method where identification of a person is done by first detecting the human face and then using a classifier and comparing it from the stored images in the database. So, it basically involves two steps. The first is face detection, and the second is face recognition.

- **Face Detection:** To identify the faces from a photograph (accurate location and size) and extract it so that it can be used by the face recognition algorithm.

- **Face Recognition:** With the face that we have extracted from the previous step, we will try to find the best possible matching face that is present in our database by applying algorithms.

For detecting faces in our system, we used the Haar-like (HAAR Cascade) feature-based algorithm, and for face recognition in our machine, we are using the following method:

6.2.2.1 Face Detection

We have used the Haar-like feature algorithm (Haar-Cascade algorithm). It is proposed as an Object-Detection algorithm, which is used to detect faces in an image or a real-time video. It uses the edge detection based features proposed by Viola and Jones in their research paper "Rapid Object Detection using a Boosted Cascade of Simple Features", published in 2001 [17]. This algorithm uses a cascade classifier that is trained on negative and positive pictures of faces. In the very start, the algorithm requires a lot of positive images (images that contain faces) and negative images (images without faces) to train the classifier. Then, we extract features from it. For this, HAAR-based features are used (shown in the below image). Each feature is a single value obtained by subtracting the sum of pixels under white rectangle from the sum of pixels under black rectangle (Figure 6.1).

6.2.2.2 Face Recognition

We have used the local binary pattern (LBP) algorithm. It is an easier face recognition algorithm and is simple to implement. Every pixel in an image is thresholded by the eight neighbor pixels, and the result is a 8-bit binary number for every pixel. It was observed that when we combine LBP with histograms of oriented gradient (HOG) descriptors, it improves the performance of detection greatly

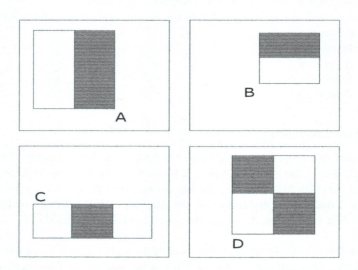

FIGURE 6.1 Type of features used in Viola-Jones algorithm (HAAR-Cascade).

on many datasets. Using LBP integrated with histograms, we may represent the facial images with a simple data vector [18–20].

6.3 HEALTH PARAMETERS AND STRESS DETECTION

The main responses of the stress response are "fight" and "flight", which are evolved in order to increase the capacity of survival of individuals or organisms. Nevertheless, longer periods of stress can result in psychological and/or somatic diseases. Anxiety is generally a bad mood due to thoughts related to anger, worry, sadness, and fear, even without the actual presence of any threats. When someone experiences anxiety frequently and at a higher intensity level without correlation to actual threat level, it can result in the person experiencing various disorders [2]. The major markers of anxiety on the human face involve the eye and its movement (eyebrows distance, blinking rate, pupil size dilation or variation), the mouth and its movement (lip deformations and shrinking), the cheeks, as well as the behavior of the head as a whole (head movements, head velocity, and movement). Other facial signs that may relate to anxiety include a tensed, stretched face, facial parlors, and twitching of eyelids [7]. A person's features, like blinking rate, aperture of eye, eyelid movement, and response and dilation/variation in pupil size, are under study. Blinking may be faked at times, but it is an automatic reaction that is found to increase with changes in emotion and depression, such as tension, stress, and anxiety. The eye blink rate is also affected by many other parameters, such as lying, and disorders like schizophrenia or depression [10].

6.3.1 Eye Blink Detection

Detection of blinking of eyes is a very important task for the system, which monitors human parameters like drowsiness while driving, physical and mental fatigue, and computer vision syndromes, such as dry eyes etc. It also helps in detecting fraud in face-recognition machines [21–24].

The human eyes play a vital role in determining the attention level of a student/individual in relation to their surroundings. At the point when an individual is exhausted, either because of absence of rest or any actual effort, they lose attention. The student can be less attentive, may lose their attention, and can be depressed or sleep deprived. Their eye movement will be different from the normal person. This eye blink rate detection can help us in determining these factors. A recent research work showed that the threshold eye blink rate for a normal person is 12–15 blinks per minute, but in a sleep-deprived subject, blinking rate will be decreased to 4–6 blink in one minute. Blinking rate is increased with changes in emotion, such as sadness, anxiety, stress, and tension [25]. (Mean observed blink is 22 blinks)

The steps used for eye blink detection are:

- Use OpenCV to load or capture video.
- Convert the frames to grayscale.
- Apply face detection using dlib.

- Get facial landmarks using dlib.
  - Load the shape predictor.
  - Map the facial landmarks.
  - Extract Cartesian coordinates and calculate EAR.

In our system, the count of eye blinks is detected by the change in EAR (eye aspect ratio) value.

6.3.2 Emotion Detection

In today's machine learning world, emotion detection is one of the most researched subjects. The ability to precisely detect and read an emotion opens up a lot of possibilities for advanced human–computer interaction. Detection of emotions can be done through human speech, body posture, and facial expression. In our project, we are using a pretrained VGG16 convolution neural network. It detects and categorizes every face based on the emotion expressed on the face of a person in one of the categories from Angry, Happy, Sad, Disgust, Fear, Surprise, and Neutral. We've used OpenCV for image processing and to recognize a face from a live webcam feed, which is then processed and fed into a trained neural network for emotion detection.

We are using the Fer2013 dataset, which we have taken from Kaggle [26,27]. Fer2013 is the dataset used in this study, which contains 32,298 images divided into training (28,709) and testing (3589) categories. It returns the probability of these possible seven emotions in the face. We will store it in our database, and if the probability of Scared or Sad face is more than 70%, we can say the person looks stressed. And if a student looks stressed for a longer course of time (a week or more), they can be assigned a mentor, or a psychiatrist session can be provided to them.

6.4 PHYSIOLOGICAL PARAMETERS AND SENSORS DESCRIPTION

We have integrated the Arduino UNO with a heart rate sensor, blood SPO2 sensor, and IR contactless temperature sensor. To serially communicate with Arduino, we have created a separate interface using "pyserial" Python library.

6.4.1 Body Temperature

According to the National Institute of Mental Health, in 2017, an "Increase in average monthly temperature by 1°F leads to a 0.48% increase in mental health issues and a 0.35% increase in suicides". Recording body temperature daily at the time of attendance can act as a possible depression marker as it strengthens the hypothesis of an inflammatory component of depression. It will also be helpful in COVID-19 prevention and symptom detection. If a student has a high fever, they may be prohibited from attending offline class as the student could have COVID-19. We have connected a buzzer that beeps when the temperature of the person is more that our set threshold.

6.4.2 Blood SpO2 Level

SpO2, also known as oxygen saturation level, is a measure of the amount of oxygen-carrying hemoglobin in the blood relative to the amount of hemoglobin not carrying oxygen. The body needs there to be a certain level of oxygen in the blood or it will not function as efficiently. Studies show that depressed people have lower oxygen level than normal in their blood. Monitoring daily SPO2 levels in students will also help to detect symptoms of COVID and anemia.

6.4.3 Heart Rate

Heart rate or pulse rate is the number of times your heart beats in one minute. Normally, the average heart rate of a person lies between 60–100 beats per minute. Faster heart rate can also be associated with diseases like heart problems, atrial fibrillation, and anemia. On average, depressed patients have a heart rate that is roughly 10–15 beats per minute higher than a person in the control group.

6.5 HARDWARE AND SENSORS USED

In our machine, we have integrated an IR contactless temperature sensor (MLX 90614), a heart beat sensor, and a blood SO2 level sensor with Arduino, which collects the health-related information of the student at the time of attendance. The Arduino will serially communicate with our attendance application with the help of Pyserial Python Library, which will activate the sensor at the time of attendance, read the parameters, and send the data to the attendance application. There, the data will be stored for the student in a separate file for the each student. From there, we can check the parameters for each student individually.

6.5.1 MLX 90614 Temperature Sensor

The MLX90614 temperature sensor is a type of contactless IR-digital temperature sensor that can measure the temperature of any object in a range of −70°C to 382.2°C from a distance. The MLX90614 sensor is a very high-accuracy IR temperature sensor. Its operation is based on a law called Stefan-Boltzmann Law, which states that all objects and living animals radite IR energy, and the amount of this radiated IR energy will be directly proportional to the temperature of that object or surface.

6.5.2 MAX 30100 Heart Rate and Blood SpO2 Sensor

The MAX30100 is a sensor that integrates the pulse oximeter together with the heart rate monitoring sensor. It uses a two-LED combination, one photo detector, and low noise processing of analog signal to measure pulse oximetry and determine the heart rate signal. The MAX30100 operating range is between 1.9 V and 3.4 V power and can be used for a longer time with high accuracy. The MAX30100 can be easily configured through our code. Its output data is stored in a 16-deep FIFO

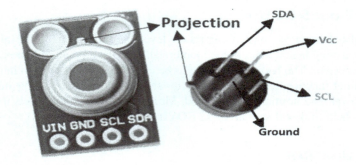

FIGURE 6.2 MLX 90614 IR contactless sensor image.

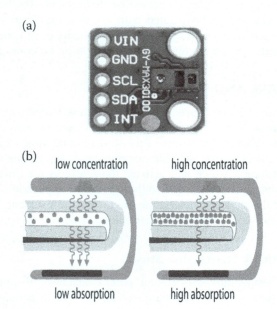

FIGURE 6.3 (a) Max3100 sensor (b) Working of Max3100 sensor.

inside the device itself. The SpO2 part of the MAX30100 consists of ambient light-cancellation and variable time filter integrated in the sensor. The SpO2 ADC is a constant time oversampling sigma delta converter with up to a 16-digit goal. The ADC yield information rate can be modified from 50 Hz to 1 kHz. The MAX30100 incorporates a restrictive discrete time channel to dismiss 50 Hz/60 Hz impedance and low-recurrence leftover encompassing commotion (Figures 6.2 and 6.3).

6.5.3 Arduino UNO

Arduino-Uno is an ATmega328P-based microcontroller. It has 14 digital pins for input and output (six are used for PWM output, six pins as analog inputs, one USB connection, and one power-jack). It also has an external button for reset. It consists of everything that is expected to help the microcontroller; we just require it to interface to a PC with a USB link. The UNO board can be powered by three different methods: by connecting it with a PC/ Laptop, by Vin pin present in the board, or by using an AC-DC adapter. In our machine, after flashing the code in Arduino, it is controlled by a Python script that initiates serial communication between Arduino and the laptop and records the data from the sensors.

6.6 SOFTWARE TOOLS

6.6.1 ARDUINO IDE

Arduino IDE (Integrated Development Environment) is used to develop the logic through which a laptop can serially communicate with Arduino. The sketch is written in Embedded C.

6.6.2 TKINTER

Tkinter is a GUI (graphical user interface) library in Python. Tkinter gives us a fast and easy way to build GUI-based applications. Tkinter also gives us a very powerful object-oriented interface to develop applications using GUI toolkits.

We have created a GUI application for our machine using Tkinter. Creating an application with a visual interface with the help of Tkinter is an easy task. We have to follow the mentioned steps:

- First, import the module in Python.
- Create a main canvas/window for our application.
- Add widgets like buttons, other windows, or tools.
- Create a logic based on loop that helps users to perform the tasks.

6.7 OUTCOMES AND RESULT

So, we have integrated all the above-mentioned details, like face recognition, emotion detection, eye blink rate, stress prediction, and health parameters, recording and combining it in the form of an application that runs on Python (Figure 6.4).

For registration of a new student, we have to enter their details, Name and ID, in the provided space and click the "TAKE IMAGE" button. It will open a new window that will detect the face of a person, and it will start taking pictures. It takes 120 pictures of the student (Figure 6.5).

6.7.1 FACE RECOGNITION

As we click on the track image button, it will start a window that will detect the face for the attendance process. The window will automatically close itself after 7 seconds (Figure 6.6).

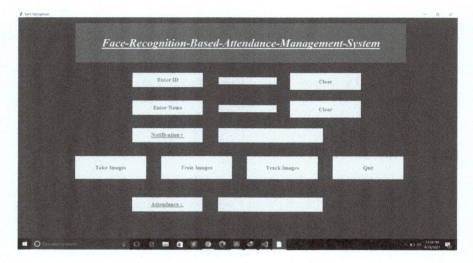

FIGURE 6.4 The image of our attendance management cum health checkup application system.

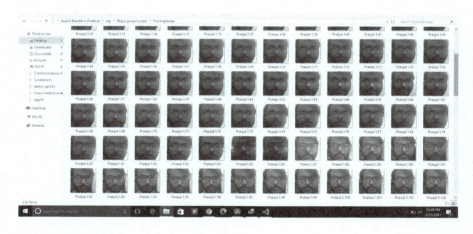

FIGURE 6.5 Stored image of student in the database.

6.7.2 Emotion Recognition and Eye-Blink Rate Detection

Once we have recognised the student, the previous window will close and a new window will open, which will monitor the eye blinks and emotion on the face of the student in the background for 15 seconds. After that, the window will close automatically, and the values of emotion and eye blinks will be recorded. Based upon that, we will calculate the stress level of the student in the background and will store the result in the csv file for the student (Figure 6.7).

Smart Attendance cum Health

FIGURE 6.6 Face recognition during attendance.

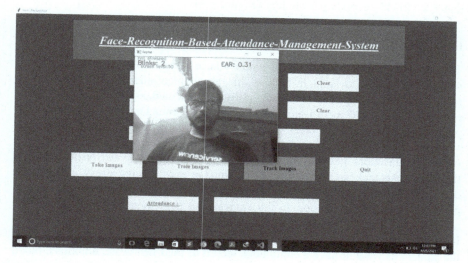

FIGURE 6.7 Eye blink and emotion recognition during attendance.

6.7.3 DATA FROM SENSORS

Once the above-mentioned step is complete, the application will begin its communication with the hardware connected to the laptop. The student has to place their fingers on the sensor, which will record the heart rate, temperature, and blood SPO2 level. Then, the recorded data will be sent to the application through the serial communication over pyserial interface between the arduino and the attendance application on our laptop. When we record the data, the student's name with their ID and the time of attendance will be on the screen of the

FIGURE 6.8 Image of sensor connected with Arduino.

application. This step takes 10–15 seconds. So, overall time for marking attendance is 40–45 seconds per student (Figure 6.8).

6.8 CONCLUSION

The aim of this project was to build an attendance system that marks the attendance of students as well as helps in their health management and checks if depression-related markers are present. We successfully implemented a system that will be able to predict stress and record emotion, and health parameters like heart rate, temperature, and blood SPO2 level from the students daily at the time of attendance. We can use some data analytics tools and also can share the results with experts like medical counsellors and doctors who can analyse and identify any health issues in the individuals and can also help students who are experiencing stress, anxiety, and depression. If we find any student who has been sad for the past week, we can provide a counselling session to the student and help them in the best possible manner i.e., either for academic issues or any personal issues. These recorded data in our database can also be used in predicting any other upcoming health issues of the student as we will have a continuous record of their heart rate, emotions, blood SPO2, and temperature.

ACKNOWLEDGEMENT

The authors are thankful to Devender Pal Singh, research scholar at MNIT Jaipur, for doing proof reading and suggesting more references for the work done.

REFERENCES

[1] R. Chandrakar, R. Raja, R. Miri, R.K. Patra, & U. Sinha, "Computer Succored Vaticination of Multi-Object Detection and Histogram Enhancement in Low

Vision", *Int. J. of Biometrics. Special Issue: Investigation of Robustness in Image Enhancement and Preprocessing Techniques for Biometrics and Computer Vision Applications*, vol. 34, 2021, pp. 1–12.

[2] G. Giannakakis, M. Pediaditis, D. Manousos, E. Kazantzaki, F. Chiarugi, P.G. Simos, K. Marias, & M. Tsiknakis, "Stress and Anxiety Detection Using Facial Cues from Videos", *Biomedical Signal Processing and Control*, vol. 31, 2017, pp. 89–101.

[3] D.J. France, R.G. Shiavi, S. Silverman, M. Silverman, & M. Wilkes, "Acoustical Properties of Speech as Indicators of Depression and Suicidal Risk", *IEEE Transactions on Biomedical Engineering*, vol. 47, no. 7, July 2000.

[4] S. Alghowinem, R. Goecke, J.F. Cohn, M. Wagner, G. Parker, & M. Breakspear, "Cross-Cultural Detection of Depression from Nonverbal Behaviour." In Automatic Face and Gesture Recognition (FG), 11th IEEE International Conference and Workshops, vol. 1, pp. 1–8. IEEE, 2015.

[5] R. Chandrakar, R. Raja, & R. Miri, "Animal Detection Based on Deep Convolutional Neural Networks with Genetic Segmentation", *Multimed Tools and Applications*, vol. 73, no. 2, 2021, pp. 1–14. 10.1007/s11042-021-11290-4

[6] G.N. Matre & S.K. Shah, "Facial Expression Detection." In IEEE International Conference on Computational Intelligence and Computing Research, December 2013.

[7] A. Pampouchidou, O. Simantiraki, C.-M. Vazacopoulos, C. Chatzaki, M. Pediaditis, A. Maridaki, K. Marias, et al., "Facial Geometry and Speech Analysis for Depression Detection." In Engineering in Medicine and Biology Society (EMBC), 39th Annual International Conference of the IEEE, pp. 1433–1436. IEEE, 2017.

[8] R. Chandrakar, R. Raja, R. Miri, U. Sinha, A.K. Kushwaha, & H. Raja, Enhanced the Moving Object Detection and Object Tracking for Traffic Surveillance Using RBF-FDLNN and CBF Algorithm. *Expert Systems with Applications*, vol. 191, 2022, 116306, ISSN: 0957-4174, 10.1016/j.eswa.2021.116306.

[9] D. Venkataraman & N.S. Parameswaran, "Extraction of Facial Features for Depression Detection among Students", *International Journal of Pure and Applied Mathematics*, vol. 118, no. 7, 2018, pp. 455–463.

[10] Y. Chouhan & R. Raja, "Face Recognition With 3D Pose Estimation Using Gabor Wavelets. International Journal of Science", *Engineering and Technology Research*, vol. 2, no. 5, 2013, ISSN: 2278-7798.

[11] S. Ghosh, S.K.P. Mohammed, N. Mogal, P.K. Nayak, & B. Champaty, "Smart Attendance System", *International Conference on Smart City and Emerging Technology (ICSCET)*, January 2018.

[12] S. Rao & K. Satoa, "An Attendance Monitoring System Using Biometrics Authentication", *International Journal of Advanced Research in Computer Science and Software Engineering*, vol. 3, no. 4, 2013.

[13] R.K. Patra, R. Raja, & T.S. Sinha, "Extraction of Geometric and Prosodic Features from Human-Gait-Speech Data for Behavioral Pattern Detection: Part II." In: Bhattacharyya S., Chaki N., Konar D., Chakraborty U., Singh C. (eds.), *Advanced Computational and Communication Paradigms. Advances in Intelligent Systems and Computing*. vol. 706, Springer, Singapore, pp. 267–278, 2018. ISBN 978-981-10-8236-8.

[14] M. Daris Femila & A. Anthony Irudhayaraj, "Biometric System." In 3rd International Conference on Electronics Computer Technology, April 2011.

[15] R. Raja, S. Kumar, S. Choudhary, & H. Dalmia, . "An Effective Contour Detection Based Image Retrieval Using Multi-Fusion Method and Neural Network", Submitted to Wireless Personal Communication, PREPRINT Version-2 available at Research Square. 2021. 10.21203/rs.3.rs-458104/v1

[16] I.H. Al Amin, E. Winarno, P.W. Adi, H. Februariyanti, M.T. Anwar, & W. Hadikurniawati, "Attendance System Based on Face Recognition System Using CNN-PCA Method and Real-time Camera." In International Seminar on Research of Information Technology and Intelligent Systems (ISRITI), December 2019.

[17] R. Raja, S. Kumar, K. RamyaLaxmi, & S. Choudhary, *Artificial Intelligence and Machine Learning in 2D/3D Medical Image Processing*. Taylor & Francis Publishing House USA, 2021. ISBN 1000337138, 9781000337136.

[18] R. Raja, K.K. Nagwanshi, S. Kumar, & K. RamyaLaxmi, *Data Mining Technologies Using Machine Learning Algorithms*. Wiley & Scrivener, 2021. ISBN No: 13: 9781119791782.

[19] R. Raja, R.K. Patra, & T.S. Sinha, "Extraction of Features from Dummy face for improving Biometrical Authentication of Human", *International Journal of Luminescence and Application*, vol. 7, no. 3-4, 2017, October–December 2017, Article 259, pp. 507–512, ISSN: 2277-6362.

[20] S. Ananth, P. Sathya, & P. Madhan Mohan, "Smart Health Monitoring System through IOT." In 2019 International Conference on Communication and Signal Processing (ICCSP), pp. 968–970, 2019. 10.1109/ICCSP.2019.8697921.

[21] Picot, A., Charbonnier, Sylvie, & Caplier, A., "Drowsiness Detection Based on vIsual Signs: Blinking Analysis Based on High Frame Rate Video." In Instrumentation and Measurement Technology Conference (I2MTC) 2010, pp. 801, 804, 3–6. IEEE, May 2010.

[22] T. Hayami, K. Matsunaga, K. Shidoji, & Y. Matsuki, "Detecting Drowsiness while Driving by Measuring Eye Movement – A Pilot Study." In Intelligent Transportation Systems, 2002. Proceedings of the IEEE 5th International Conference, pp. 156, 161, 2002.

[23] A.I. Siam, A.E. Abouelazm, N.A. El-Bahnasawy, G. El Banby, & F.E.A. El-Samie, "Smart Health Monitoring System based on IoT and Cloud Computing", ICEEM2019-Special Issue, Article 30, vol. 28, pp. 37–42, 2019.

[24] S. Harati, A. Crowell, H. Mayberg, J. Kong, & S. Nemati, "Discriminating Clinical Phases of Recovery from Major Depressive Disorder Using the Dynamics of Facial Expression." In Engineering in Medicine and Biology Society (EMBC), 38th Annual International Conference, pp. 2254–2257. IEEE, 2016.

[25] L. Tiwari, R. Raja, V. Sharma, & R. Miri, "Fuzzy Inference System for Efficient Lung Cancer Detection." In: Gupta M., Konar D., Bhattacharyya S., Biswas S. (eds.), *Computer Vision and Machine Intelligence in Medical Image Analysis. Advances in Intelligent Systems and Computing*, 992. Springer, Singapore, 2020. Online ISBN978-981-13-8798-2.

[26] L. Tiwari, Raja, V., Awasthi, R., Miri, G.R., Sinha, M.H. Alkinani, & K. Polat, "Detection of Lung Nodule and Cancer Using Novel Mask-3 FCM and TWEDLNN Algorithms", *Measurement*. vol. 172, 2021, p. 108882, ISSN: 0263-2241. 10.1016/j.measurement. 2020.108882.

[27] L. Tiwari, R. Raja, V. Sharma, & R. Miri, "Adaptive Neuro Fuzzy Inference System Based Fusion of Medical Image", *International Journal of Research in Electronics and Computer Engineering*, vol. 7, no. 2, 2020, pp. 2086–2091, ISSN: 2393-9028 (PRINT) |ISSN: 2348-2281 (ONLINE).

7 Oral Histopathological Photomicrograph Classification Using Deep Learning

Rajashekhargouda C. Patil
Department of Electronics and Communication Engineering,
Don Bosco Institute of Technology, Bengaluru, India

P. K. Mahesh
Department of Electronics and Communication Engineering,
Academy for Technical and Management Excellence College
of Engineering, Mysore, India

CONTENTS

7.1 Introduction 85
7.2 Related Work 86
7.3 Present Diagnosing Method for Oral Cancer 87
7.4 Materials and Methods 89
    7.4.1 Deep-Learning Combined with SVM Approach 90
    7.4.2 Transfer Learning of the Deep-Learning Model's Approach 92
    7.4.3 Fusion of the Results Obtained from Transfer Learning and SVM Process 95
7.5 Results and Discussions 96
    7.5.1 Comparison of the Results with Other Related Works 98
7.6 Conclusion 103
References 103

7.1 INTRODUCTION

Oral cancer is a life-threatening disease that originates in the oral cavity. This occupies the third position among all types of cancers (Coelho 2012). The prediction on oral cancer incidence estimates that the disease will affect more than 100,000 (Coelho 2012).

Presently diagnosis of oral cancer is through the analysis of the histopathological photomicrographs that are obtained on the biopsied tissue after staining them with the Hematoxylin and Eosin (H&E) stain. H&E stain is the most widely used stain in pathology laboratories as it helps diagnose and stage most of the malignancies. This stain colours the acidic part of the tissue (nucleic acids) in the bluish shade and the basic part of the tissue in the reddish shade. The analysis through histopathology is done manually and allows the errors or misclassifications to creep in since it is purely based on the experience and knowledge of the pathologists (Woolgar et al. 2011). In the view of eliminating this error in pathology, we are proposing an automation algorithm by employing the deep learning technique.

Deep learning is the technology of generating the features through the inspection of images and classification of the images depending on the content of the image. This field has evolved in the very recent past. This technology uses the convolution neural network concept.

7.2 RELATED WORK

In this section, we will present the survey of some related research carried out. We restrict our survey of related work to oral cancer detection using artificial intelligence methods.

A. Chodorowski et al. (1999) used many colour representations i.e. Reg-Green Blue(RGB), Irg, Hue-Saturation-Intensity (HIS), I1 I2 I3 and La*b*) to detect oral lichenoid reactions and oral leukoplakia. Features extracted for the classification consist of the mean colour difference between the normal and the abnormal sections. The classification was done with the help of Gaussian quadratic, Fisher's linear discriminant, Multilayer Perceptron, and K-nearest neighbor (KNN) classifiers. On the dataset of 70 images of oral lichenoid reactions and 20 images of homogenous oral leukoplakia, this study achieved an accuracy of 94.6% and 70.0%, respectively, of correct classification.

Khosravi, Pegah et al. (2018) (Chandrakar et al. 2020) proposed the classification process to separate the two subtypes of lung cancer, four biomarkers of bladder cancer, and five biomarkers of breast cancer. The study employed a basic convolution neural network (CNN) architecture, Google's Inceptions with three training strategies, and an ensemble of two state-of-the-art algorithms, Inception and ResNet. Training strategies included training the last layer of Google's Inceptions, training the network from scratch, and fine-tuning the parameters using two pre-trained versions of Google's Inception architectures, Inception-V1 and Inception-V3. The results obtained accuracies of 100%, 92%, 95%, and 69% for discrimination of various cancer tissues, subtypes, biomarkers, and scores, respectively.

Maisun Mohamed Alzorgani and Hassan Ugail (2018) performed the classification of 1424 histopathological photomicrographs related to head and neck tumours, among which 1184 images were of squamous cell carcinoma and 240 images were of normal tissue. This study did the transfer learning of Resnet 50, Resnet 101, GoogLeNet, and Inception V3 with the above-mentioned images. The accuracies obtained were 98.95% for Resnet50, 97.89% for ResNet-101, 97.19% for GoogleNet, and 94.04% for Inception-v3.

M. Muthu Rama Krishnan et al. (2012) extracted granular structures that had self-similar patterns at different scales, termed "texture", as the features of H&E stained images using higher-order spectra (HOS), local binary pattern (LBP), and laws texture energy (LTE) and that applied to the fuzzy classifier. They obtained the accuracy of 95.7% with the sensitivity of 94.5% and specificity of 98.8%, respectively.

B. Kieffer et al. (2017) applied the Kimia Path24 dataset, which consists of 27,055 histopathology training patches in 24 tissue texture classes along with 1325 test patches for evaluation on pre-trained model Visual Geometry Group (VGG) 16 & inception models and the fine-tuned VGG 16 & inception models. The authors concluded that fine-tuning VGG 16 will not yield appreciable better results when fine-tuned, whereas inception on fine-tuning will provide much better results. The total accuracy that was the product of patch-to-scan accuracy and whole-scan accuracy got improved for the transfer learned Inception V3 model from 50.54% to 56.98%.

Das et al. (2015) proposed a novel method to identify the keratinization and keratin pearls in the H&E stained images. The method included colour space transform in *YDbDr* channel, enhancement of keratinized area in most significant bit (MSB) plane of *Db* component, and segmentation of keratinized area using the Chan–Vese model (Getreuer 2012). This study was able to achieve an accuracy of 95.08%.

Rahman et al. (2018) came up with an algorithm for classifying cancer and normal pathological images. They used the textual features obtained from the Gray Level Coherence Matrix (GLCM) method and, upon applying the Principal Component Analysis (PCA) on the feature set, the classification was done using the SVM classifier. This study achieved an accuracy of 100%, with both the specificity and sensitivity reaching 100%. This study focused on the abnormalities in the epithelium layer and hence cropped only the epithelium layer from the images, which consisted of the epithelium layer along with the unwanted connective tissue and other background parts. Also, this study was dependent on the ROI suggested by the pathologist.

Noroozi and Zakerolhosseini (2016) proposed an automation method for the classification of squamous cell carcinoma of the skin. This study carried out the segmentation of the epidermis along with cornified layer removal. Then, the epidermis axis was specified using the paths in its skeleton, and the granular layer was removed via connected components analysis. Features for the classifier were the intensity profiles extracted from lines perpendicular to the epidermis axis. The classification was carried out using the SVM technique. This study obtained the sensitivity, precision, specificity, and accuracy of 84.6%, 84.6%, 81.8%, and 83.3%, respectively.

Das et al. (2018) proposed a two-stage procedure for the detection of oral cancer in histology images. In the first stage, segmentation was carried out using the convolution neural networks (CNN) approach, and in the second, keratin pearls were detected in the segmented region using texture-based (Gabor filter) trained random forests. This study achieved a detection accuracy of 96.88%.

7.3 PRESENT DIAGNOSING METHOD FOR ORAL CANCER

Diagnosing through histopathology images is the only method approved by all the Health Departments throughout the world as the diagnosing method in cancer detection and staging and hence is referred to as the "gold standard". In this method, once the biopsy specimen is ready after staining with the H&E stain, the pathologist places the slide under the microscope for visual inspection.

In the microscope, we have the lenses present at two ends of the microscope tube. The lens that is in contact or nearest to the eye is called the eyepiece. Usually, the eyepiece lens is of 10x magnification. The other lens that is present at the other end of the tube is called the objective. The objective lens will be three in number and usually will be of 10x, 20x, and 40x magnification. Thus, a combination of the 10x eyepiece and the 10x objective lens together will yield a magnification of 100x, the combination of the 10x eyepiece with the 20x objective lens will provide the magnification of 200x, and similarly, we will obtain 400x magnification by combining the 10x eyepiece with the 40x objective lens.

In 400x magnification, the observation that can be made is that of the shape and size of the nucleus in the cell (Das et al. 2018). If the nucleus is covering most of the cell and if the nucleus is of irregular shape, then the sample is of cancer affected tissue. If the nucleus is small compared to the size of the cell and is a regular shape, then the sample is of benign tissue. Figure 7.1 shows the difference between the normal cell and the cancer cell under 400x magnification, where the nuclei are dark purple.

The information obtained from 100x magnification is the pattern of the epithelium (Sudhakara 2016) and the presence of the keratin pearls. If the sample is of malignant tissue, the pattern of the epithelium will be haphazard. If the tissue is benign, then the epithelium will be in a comb structure.

If the sample is of benign tissue, then the epithelium will be present only at the border or the surface area, whereas if the sample is of malignant tissue, then the epithelium will be found invading the second layer, which is the connective tissue. Figure 7.2 shows the H&E stained photomicrograph of a benign tissue under 100x magnification, where the epithelium layer will have a regular comb-like pattern, and Figure 7.3 depicts the H&E stained photomicrograph of malignant tissue under 100x magnification, where the invasion of the epithelium into the connective tissue can be observed as indicated by the dark blue arrow.

Another important observation that can be done in 100x magnification is the presence of Keratin pearls (Caliaperoumal 2016). Keratin is a protein and can be observed in the samples of malignant tissue in the form of a spiral structure. The presence of Keratin pearl increases the confidence to classify the image as that of malignant tissue, but the absence of Keratin pearls does not provide 100% probability that the image is of

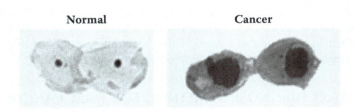

FIGURE 7.1 Cancer cells vs normal cells.

Courtesy: https://sphweb.bumc.bu.edu/otlt/mph-modules/ph/ph709_cancer/ph709_cancer7.html

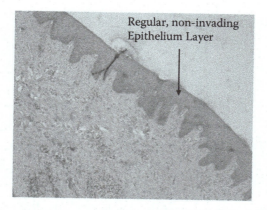

FIGURE 7.2 H&E stained biopsy sample of normal tissue depicting regular non-invading Epithelium.

Source: Photograph collected from RR Dental Hospital.

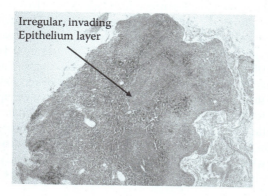

FIGURE 7.3 H&E stained biopsy sample of malignant tissue depicting irregular invading Epithelium.

Source: Photograph collected from RR Dental Hospital.

benign tissue as some malignant images taken in the initial stage of cancer do not have Keratin pearls. Figure 7.4 depicts the presence of Keratin pearls in the sample taken from malignant tissue, highlighted by the black arrow.

7.4 MATERIALS AND METHODS

Our experimental analysis consists of a total of 21 benign images and 98 malignant images, among which 10% are for testing and the remaining are used for training purposes. All the images used are of H&E stained images obtained at 10x magnification level with respect to the objective lens specification. In our proposed algorithm, we have used ResNet 50 (Kaiming et al. 2015),

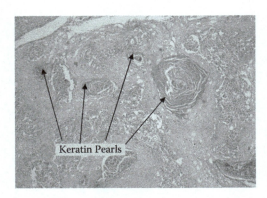

FIGURE 7.4 H&E stained biopsy sample of malignant tissue depicting Keratin pearls.
Source: Photograph collected from RR Dental Hospital.

ResNet 101 (Kaiming et al. 2015), VGG 16 Net (Simonyan and Zisserman 2014), VGG 19 Net (Simonyan and Zisserman 2014), AlexNet (Krizhevsky et al. 2012), GoogLeNet (Szegedy et al. 2015), and Inceptionv3 Net (Szegedy et al. 2016).

These models can be used in two different ways. One approach is through the transfer learning process in which we will use all the models except the last three layers. As the models are built to classify the image into 1000 different classes, in our study we have only two classes and hence have to alter them, which is called the "transfer learning" technique. The other approach is to get the features and to provide those features to the SVM classifier. The helplessness we observed while using SVM is that it will just give the class as the result, like malignant or benign. We will not be able to get the probability of classification through SVM, which is available through the first approach. We have used both approaches in our work and have fused both the results with appropriate weights.

Before proceeding further, let us first understand the notations. The term "labelled" is the class given by the pathologist, which is known to us in advance. The term "classified" or "predicted" is the output of the program.

7.4.1 Deep-Learning Combined with SVM Approach

Here we will extract the features through the deep learning process and will provide these extracted features to the SVM classifier, which will train itself on these features and will classify a test image based on the trained knowledge. Table 7.1 provides the layer names from which the features can be extracted for each of the models we have employed in our work.

As discussed earlier, this approach of using SVM to classify the extracted features will provide us with the result of the test image, not in probability, but in class. Once the SVM is trained, it will always result in the same class for a particular test image if the image is given to be classified any number of times, but this is not the case with the model. In one execution, the model may result in

TABLE 7.1

Feature extraction layers

Model	Layer Name
Alexnet	fc8
VGG16	fc8
VGG19	fc8
GoogLeNet	loss3-classifier
Inception V3	predictions
Resnet 50	fc1000
Resnet 101	fc1000

some features, whereas in the next execution, the features provided by the model may be a bit different from that of the previous run for the same set of training images. Models do not learn features from only one image but will generalise the features from a batch of images belonging to a particular class. The change in the run to run will be for those images that are in the doubtful class. We have used the following method to get the end conclusion to form the Model-SVM approach.

1. Each model is executed 20 times on the complete image set. Let the variable that denotes the run number of a particular model be 'r'. Let the models be denoted by 'm' and the images be denoted by 'i', which includes both benign and malignant images. Since each image is given 20 times as the input to a particular model, we will get 20 results, which will be either the correct class or a wrong class.
2. In each of the 20 runs, the output class ($C_{m,r,i}$) is converted to a number ($L_{m,r,i}$) as per Expression 7.1.

$$If\ (C_{m,r,i} = \text{``Normal''})\ then\ L_{m,r,i} = -1),\ else\ (L_{m,r,i} = 1) \quad (7.1)$$

3. Compute the gain of the image from all 20 runs, as given in Expression 7.2 below.

$$L_{m,i} = \sum_{r=1}^{20} L_{m,r,i} * 0.05 \quad (7.2)$$

4. Compute the output class for the image by applying the threshold Φ_1 to the obtained $L_{m,i}$ as per Expression 7.3. 'S' in "CS" indicates the SVM process.

$$If\ (L_{m,i} > \Phi_1\ then\ CS_{m,i} = \text{``Cancer''}),$$
$$elseif\ (L_{m,i} < -\Phi_1\ then\ CS_{m,i} = \text{``Normal''}) \quad (7.3)$$
$$else\ (CS_{m,i} = \text{``Doubt''})$$

5. The threshold Φ_1 is fixed to an optimum value for a particular model for the SVM section such that the misclassification error is as small as possible and simultaneously the "doubt" class also contains as few samples as possible.
6. Compute the respective weightage for doubt ($WS_{m,D}$), true positive ($WS_{m,TP}$), true negative ($WS_{m,TN}$), false positive ($WS_{m,FP}$), and false negative ($WS_{m,FN}$) as per Expression 7.4, given below. 'S' in "WS" indicates that the value is from the SVM process.

$$WS_{m,D} = \frac{Number\ of\ samples\ in\ Doubt\ Section}{Total\ number\ of\ samples}$$
$$WS_{m,TP} = \frac{Number\ of\ samples\ in\ TP\ Section}{Total\ number\ of\ samples\ labelled\ as\ Cancer}$$
$$WS_{m,TN} = \frac{Number\ of\ samples\ in\ TN\ Section}{Total\ number\ of\ samples\ labelled\ as\ Normal} \quad (7.4)$$
$$WS_{m,FN} = \frac{Number\ of\ samples\ in\ FN\ Section}{Total\ number\ of\ samples\ labelled\ as\ Normal}$$
$$WS_{m,FP} = \frac{Number\ of\ samples\ in\ FP\ Section}{Total\ number\ of\ samples\ labelled\ as\ Cancer}$$

Where FP: False Positive (Total Count of Labelled benign, predicted malignant)
FN: False Negative (Total Count of Labelled malignant, predicted benign)
TP: True Positive (Total Count of Labelled malignant, predicted malignant)
TN: True Negative (Total Count of Labelled benign, predicted benign)

7. The next step is to rate the model for the SVM process for all three classes (normal, doubt, cancer), depending on the weights computed in step 6. The weightage $WS_{m,TP}$ is to be considered positive as true positive is one of the expected outputs. The other expected output is true negative, and hence $WS_{m,TN}$ is also considered positive. The error is when labelled normal samples predicted as cancer or when the sample labelled as cancer is predicted as normal; hence, these two weights ($WS_{m,FP}$ and $WS_{m,FN}$) are considered with a double negative penalty. The doubt rate should be as small as possible and hence weighted with a single penalty. The model rate for SVM (MRS) is computed for all three classes as given by Expression 7.5.

$$MRS_{m,Normal} = WS_{m,TN} - 2 * WS_{m,FN}$$
$$MRS_{m,Cancer} = WS_{m,TP} - 2 * WS_{m,FP} \quad (7.5)$$
$$MRS_{m,Doubt} = 1 - WS_{FP}$$

7.4.2 Transfer Learning of the Deep-Learning Model's Approach

In this 'Models-TL' method, we will truncate the model by the last three layers and will replace them with a fully connected layer, followed by the

Softmax layer, followed by the classification layer as the output layer with only two nodes. Table 7.2 lists the names of the layers removed, with the previous layer (4th from last) name joined to the fully connected layer for each of the models used in our work. This truncation and replacement of layers was done to reduce the number of output nodes to two (nodeN and nodeC) as most of the models have 1000 output nodes to classify the image into 100 different classes. Transfer learning is the process of transferring the learned knowledge of classifying in one domain to the classification in another domain.

In our case, the models that have learned to classify 1000 different objects visible in our day-to-day surrounding pictures like animals, human beings, vehicles etc. are transferring their learned knowledge to classify cancer and normal H&E stained photomicrographs, which we do not see in the normal photographs

1. Each model is executed 20 times on the complete image set.
2. Fetch the maximum value for both the nodes (nodeN & nodeC) from all the runs. Let us denote these terms as $nodeNmax_{m,i}$ and $nodeCmax_{m,i}$.
3. Fetch the median of all nodeN (from all 20 runs). Let us denote this term as $nodeNmed_{m,i}$. Similarly, fetch the median of all nodeC (from all 20 runs). Let us denote this term as $nodeCmed_{m,i}$. Median is the value that occupies the center (middle) position when the data are arranged in ascending order. If the data are even, then the median is the average of the two positions at the center.
4. Compute the median of all $nodeNmed_{m,i}$ of the subset where the label is normal. Let us denote this value as $nodeNmed_m$. Similarly, calculate the median of all $nodeCmed_{m,i}$ of the images whose label is cancer. Let us denote this value as $nodeCmed_m$. Please note that the values $nodeNmed_m$ and $nodeCmed_m$ are calculated only for the training samples. Testing samples use the values of $nodeCmed_m$ and $nodeCmed_m$ of the training itself.
5. Let us define the threshold Φ_2. For every image of a particular model, predict the class for TL process as per the computation given in Expression 7.6 below.

$$\begin{aligned} CT_{m,i} &= \text{"Normal"} \; if \; (nodeNmax_{m,i} > (nodeNmed_m - \Phi_2)) \\ &= \text{"Cancer"} \; if \; (nodeCmax_{m,i} > (nodeCmed_m - \Phi_2)) \\ &\quad else \; \text{"Doubt"} \end{aligned} \quad (7.6)$$

6. Compute the respective weightage for doubt ($WT_{m,D}$), true positive ($WT_{m,TP}$), true negative ($WT_{m,TN}$), false positive ($WT_{m,FP}$), and false negative ($WT_{m,FN}$) as per Expression 7.7 given below. 'T' in "WT" indicates that the value is of the transfer learning process.

TABLE 7.2
Layer modification in Model-TL approach

Model	Resnet 50	Resnet 101	Googlenet	VGG 16 / 19 / Alexnet	Inceptionv3
Fourth Layer from last before Truncation	'avg_pool'	'pool5'	'pool5-drop_7x7_s1'	'drop7'	'avg_pool'
Last three Layers Removed	'avg_pool', 'fc1000_softmax', 'ClassificationLayer_fc1000'	'fc1000', 'prob', 'ClassificationLayer_predictions'	'loss3-classifier', 'prob', 'output'	'fc8', 'prob', 'output'	'predictions', 'predictions_softmax', 'ClassificationLayer_predictions'

Oral Photomicrograph Classification

$$WT_{m,D} = \frac{Number\ of\ samples\ in\ Doubt\ Section}{Total\ number\ of\ samples}$$

$$WT_{m,TP} = \frac{Number\ of\ samples\ in\ TP\ Section}{Total\ number\ of\ samples\ labelled\ as\ Cancer}$$

$$WT_{m,TN} = \frac{Number\ of\ samples\ in\ TN\ Section}{Total\ number\ of\ samples\ labelled\ as\ Normal} \quad (7.7)$$

$$WT_{m,FN} = \frac{Number\ of\ samples\ in\ FN\ Section}{Total\ number\ of\ samples\ labelled\ as\ Normal}$$

$$WT_{m,FP} = \frac{Number\ of\ samples\ in\ FP\ Section}{Total\ number\ of\ samples\ labelled\ as\ Cancer}$$

Where FP: False Positive (Total Count of Labelled benign, predicted malignant)
FN: False Negative (Total Count of Labelled malignant, predicted benign)
TP: True Positive (Total Count of Labelled malignant, predicted malignant)
TN: True Negative (Total Count of Labelled benign, predicted benign)

7. The next step is to rate the model for TL process for all three classes (normal, doubt, cancer), depending on the weights computed in step 6. The weightage $WT_{m,TP}$ is to be considered positive as true positive is one of the expected output. The other expected output is true negative, and hence $WT_{m,TN}$ is also considered positive. The error is when labelled normal samples predicted as cancer or when the sample labelled as cancer is predicted as normal; hence, these two weights ($WT_{m,FP}$ and $WT_{m,FN}$) are considered with a double negative penalty. The doubt rate should be as less as possible and hence weighted with a single penalty. The model rate for TL (MRT) is computed for all three classes as given by Expression 7.8.

$$MRT_{m,Normal} = WS_{m,TN} - 2 * WS_{m,FN}$$
$$MRT_{m,Cancer} = WS_{m,TP} - 2 * WS_{m,FP} \quad (7.8)$$
$$MRT_{m,Doubt} = 1 - WS_{FP}$$

7.4.3 Fusion of the Results Obtained from Transfer Learning and SVM Process

Now, as we have the class predictions from both processes (TL and SVM) of all models for all images, we will use these predictions to provide the final prediction as per the steps listed below.

1. Assign numerical values for the classes obtained from the two processes of all the models for all the images. The assignment method is given below. 'X' can be 'S' or 'T' i.e. of SVM or TL process. Let us call the assigned value 'Y' with the appropriate suffix.

$$\text{If } CX_{m,i} = \text{``Normal''} \text{ then } YX_{m,i} = 1$$
$$\text{elseif } CX_{m,i} = \text{``Cancer''} \text{ then } YX_{m,i} = -1 \quad (7.9)$$
$$\text{If } CX_{m,i} = \text{``Doubt''} \text{ then } YX_{m,i} = 0$$

2. Multiply the assigned value '$YX_{m,i}$' with the respective model weightage. For e.g. if the considered 5th sample has obtained class as normal from the SVM process of model 3, then the output of this step will be $1*WS_{3,5}$.
3. Repeat step 2 for both values of 'X', all seven values of 'm', and all samples 'i'.
4. After completion of step 3, we are left with 14 values (7 from SVM and 7 from TL). Take the average of all the 14 values for a particular sample, which will yield a value. Let us call this average value $Favg_i$.
5. Assign class to the $Favg_i$ by fixing the threshold Φ_3. We have observed in our work that $Favg_i$ will be positive for normal samples and negative for cancer samples. This implies that $\Phi_3 = 0$ classifies the sample into normal and cancer class. We can bring in the doubt class by modifying the bifurcation line into a window whose one end on the normal side will have the value Φ_{3N}, whereas the other end of the window on the cancer side will have the value Φ_{3C}. All the samples that fall in this window (Φ_{3N} to Φ_{3C}) will be classified as doubt.

The point here not to be forgotten is that the median value calculated for a model in the transfer learning process will be only for the training images. The testing image will use this training median value itself along with its maximum value and the threshold Φ_2 in the computation of the class (step 5 of the transfer learning process).

7.5 RESULTS AND DISCUSSIONS

In this section, the confusion matrix parameters obtained through Model-SVM process and Model-TL process are presented. Also, the model weights computed through the confusion matrix parameters and the model rates computed on model weights for both the processes are listed out.

Table 7.3 gives the doubt rate, true positive rate, true negative rate, false positive rate, and false negative rate of the employed modes for the Model-SVM process. Table 7.4 provides the list of weights calculated through Expression 7.4 for the Model-SVM process of all seven models. Table 7.5 lists out the model rates of normal, doubt, and cancer class for all used models computed through Expression 7.5. Figure 7.5 shows the graphical representation of Table 7.5.

Table 7.6 gives the doubt rate, true positive rate, true negative rate, false positive rate, and false negative rate of the employed modes for the Model-TL process.

Table 7.7 provides the list of weights calculated through Expression 7.7 for the Model-TL process of all seven models. Table 7.8 lists the model rates of normal, doubt, and cancer class for all used models computed through Expression 7.5. Figure 7.6 shows the graphical representation of Table 7.8.

TABLE 7.3
Confusion matrix parameter rates for Model-SVM process

	AlexnetSVM	VGG19SVM	VGG16SVM	GooglenetSVM
Doubt Rate	31.73%	27.88%	18.27%	100.00%
TP Rate	64.77%	67.05%	78.41%	0.00%
FN Rate	2.27%	1.14%	0.00%	0.00%
FP Rate	0.00%	0.00%	0.00%	0.00%
TN Rate	75.00%	93.75%	100.00%	0.00%
	Resnet101SVM	Resnet50SVM	InceptionV3SVM	
Doubt Rate	26.92%	30.77%	32.69%	
TP Rate	68.18%	67.05%	63.64%	
FN Rate	0.00%	0.00%	0.00%	
FP Rate	0.00%	0.00%	0.00%	
TN Rate	100.00%	81.25%	87.50%	

TABLE 7.4
Weights assigned through Expression 7.4

Weights	AlexnetSVM	VGG19SVM	VGG16SVM	GooglenetSVM
Doubt	0.32	0.28	0.18	1.00
TP	0.65	0.67	0.78	0.00
FN	0.13	0.06	0.00	0.00
FP	0.00	0.00	0.00	0.00
TN	0.75	0.94	1.00	0.00
Weights	Resnet101SVM	Resnet50SVM	InceptionV3SVM	
Doubt	0.27	0.31	0.33	
TP	0.68	0.67	0.64	
FN	0.00	0.00	0.00	
FP	0.00	0.00	0.00	
TN	1.00	0.81	0.88	

Now, as the training procedure is completed, providing the model rates, we will apply the obtained model rates for normal, doubt, and cancer classes to the predicted output and will take the sum for final classification. Table 7.9 gives the predictions of each model in terms of classes.

S, T, N, D, and C represent SVM Process, TL Process, Normal Prediction, Doubt Prediction, and Cancer Prediction, respectively. Table 7.10 provides the product of the classes and the respective class model rate.

TABLE 7.5
Model rates for three classes assigned through Expression 7.5

Model Rate	AlexnetSVM	VGG19SVM	VGG16SVM	GooglenetSVM
Normal	0.50	0.81	1.00	0.00
Doubt	0.68	0.72	0.82	0.00
Cancer	0.65	0.67	0.78	0.00
Model Rate	Resnet101SVM	Resnet50SVM	InceptionV3SVM	
Normal	1.00	0.81	0.88	
Doubt	0.73	0.69	0.67	
Cancer	0.68	0.67	0.64	

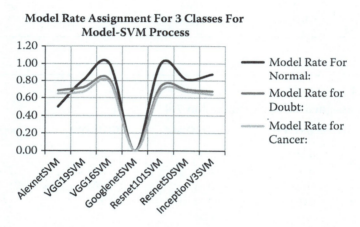

FIGURE 7.5 Graphical representation of Table 7.5.

As we can see in Table 7.10, all the test samples are classified correctly. Also when we applied the same classifying technique to the training samples, we found the classification accuracy to be 100%. Thus, with the proof through experimental analysis, we are proposing a robust H&E stained photomicrograph classification technique.

It would not make much sense to provide the graph of every run of every model used as there are 20 runs for each of the seven models. Hence, for demonstration purposes, one graph of the training and validation is shown in Figure 7.7.

7.5.1 Comparison of the Results with Other Related Works

Rahman et al. (2018) has also achieved the result of 100% accuracy, 100% sensitivity, and 100% specificity. Their data set consists of 134 normal images and 135 cancer images. There is a small difference in the number of cancer images but a huge difference pops up in the number of normal images when

TABLE 7.6
Confusion matrix parameter rates for Model-TL process

	AlexnetTL	VGG19TL	VGG16TL	GooglenetTL
Doubt Rate	0.00%	0.00%	1.92%	9.62%
TP Rate	100.00%	100.00%	97.73%	90.91%
FN Rate	0.00%	0.00%	0.00%	0.00%
FP Rate	25.00%	6.25%	6.25%	6.25%
TN Rate	75.00%	93.75%	93.75%	81.25%
	Resnet101TL	Resnet50TL	InceptionV3TL	
Doubt Rate	0.00%	0.00%	12.50%	
TP Rate	100.00%	100.00%	86.36%	
FN Rate	0.00%	0.00%	0.00%	
FP Rate	6.25%	0.00%	0.00%	
TN Rate	93.75%	100.00%	93.75%	

TABLE 7.7
Weights assignment for Model-TL process through Expression 7.7

Weights	AlexnetTL	VGG19TL	VGG16TL	GooglenetTL
Doubt	0.00	0.00	0.02	0.10
TP	1.00	1.00	0.98	0.91
FN	0.00	0.00	0.00	0.00
FP	0.05	0.01	0.01	0.01
TN	0.75	0.94	0.94	0.81
Weights	Resnet101TL	Resnet50TL	InceptionV3TL	
Doubt	0.00	0.00	0.13	
TP	1.00	1.00	0.86	
FN	0.00	0.00	0.00	
FP	0.01	0.00	0.00	
TN	0.94	1.00	0.94	

compared to our data set. The images that they have used are of 400x magnification, and our images were of 100x magnification.

The methods used in both the researches are completely different. They have used the textural features obtained from Gray-Level Co-occurrence Matrix (GLCM features) extracted only from the specific part of the image that is marked by the pathologist. In our case, we have not carried out any such region of interest (ROI) extraction; hence, our work is not dependent on the knowledge of ROI extraction, even in training.

TABLE 7.8
Model rates for three classes of Model-TL process through Expression 7.8

Model Rate	AlexnetTL	VGG19TL	VGG16TL	GooglenetTL
Normal	0.75	0.94	0.94	0.81
Doubt	1.00	1.00	0.98	0.90
Cancer	0.91	0.98	0.95	0.89
Model Rate	Resnet101TL	Resnet50TL	InceptionV3TL	
Normal	0.94	1.00	0.94	
Doubt	1.00	1.00	0.88	
Cancer	0.98	1.00	0.86	

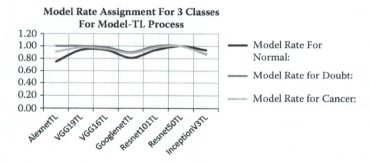

FIGURE 7.6 Graphical representation of Table 7.8.

Coming to the testing part, the paper does not specify whether the textural features extracted from the testing images are also from only within the ROI or of the complete image. With the logical reasoning that if the textural features are extracted from the complete testing image, then the textural features extracted from the training and testing images will be different from each other. This logic drives us to conclude that the textural features extracted from the testing image are also only from the ROI marked up by the pathologist.

And if the pathologist's help is needed in the testing of the images, then it makes very little sense to automatize the detection procedure in their method. The aim of automatizing is to make the detection process as independent of the pathologist as possible.

Hence, we feel that, even on having the same accuracy, sensitivity, and specificity of 100%, our work is superior as it does not need the help of the pathologist in the testing process.

Even though many studies have been published to classify the pathological H&E stained photomicrographs into cancer and normal class, apart from the one mentioned work (Rahman et al. 2018), none of the individual researchers or any group, to the best of our knowledge, has provided 100% accuracy, sensitivity, and specificity.

TABLE 7.9
Model output in terms of class for the test images

Image	Alexnet		VGG19		VGG16		Googlenet		Resnet101		Resnet50		InceptionV3	
	S	T	S	T	S	T	S	T	S	T	S	T	S	T
C1	C	C	C	D	C	C	C	D	C	C	C	C	D	C
C2	C	C	D	C	D	C	D	D	C	C	C	C	C	C
C3	C	C	C	C	C	C	C	D	C	C	C	C	C	C
C4	C	C	D	D	C	D	D	D	C	C	C	C	C	C
C5	C	C	C	C	C	C	C	C	C	C	C	C	C	C
C6	C	C	C	C	C	C	C	D	C	C	C	D	C	C
C7	C	D	C	C	C	C	C	D	C	C	C	C	C	C
C8	C	D	D	D	D	D	D	D	C	C	D	D	D	D
C9	C	N	D	D	D	D	D	D	C	C	D	D	D	C
C10	C	N	D	D	D	D	D	D	C	D	D	D	D	D
N1	C	N	D	N	N	N	D	D	C	N	D	N	D	N
N2	C	D	D	N	D	D	D	D	D	N	D	D	D	N
N3	C	N	D	N	D	N	D	D	C	N	D	D	D	N
N4	N	D	D	N	D	N	D	D	D	N	D	D	D	N
N5	N	N	D	N	C	D	D	D	C	N	D	D	D	D

TABLE 7.10
Assignment of final class

	C1	C2	C3	C4	C5	C6	C7	C8	C9	C10	N1	N2	N3	N4	N5
Alexnet TL	-0.91	-0.91	-0.91	-0.91	-0.91	-0.91	-0.91	-0.91	-0.91	-0.91	-0.91	-0.91	-0.91	0.75	0.75
Alexnet SVM	-0.77	-0.77	-0.77	-0.77	-0.77	-0.77	-0.77	0.00	0.00	0.75	0.75	0.00	0.75	0.00	0.75
VGG19 TL	-0.98	0.00	-0.98	-0.98	-0.98	-0.98	-0.98	-0.98	0.00	0.00	0.00	0.00	0.00	0.00	0.00
VGG19 SVM	0.00	-0.72	-0.72	0.00	-0.72	-0.72	-0.72	-0.72	0.63	0.00	0.63	0.63	0.63	0.63	0.63
VGG16 TL	-0.95	0.00	-0.95	-0.95	-0.95	-0.95	-0.95	-0.95	0.00	0.00	0.00	0.94	0.94	0.00	0.00
VGG16 SVM	-0.88	-0.88	-0.88	-0.88	-0.88	-0.88	-0.88	-0.88	0.00	0.00	0.88	0.00	0.88	0.88	0.00
Googlenet TL	-0.89	0.00	-0.89	0.00	-0.89	-0.89	-0.89	0.00	-0.86	0.00	0.00	0.00	0.00	0.00	0.00
Googlenet SVM	0.00	0.00	0.00	0.00	0.00	0.00	-0.05	0.00	0.00	0.00	0.00	0.00	0.00	0.00	0.00
Resnet101 TL	-0.98	-0.98	-0.98	-0.98	-0.98	-0.98	-0.98	-0.98	-0.98	-0.98	-0.98	-0.98	0.00	-0.98	-0.98
Resnet101 SVM	-0.81	-0.81	-0.81	-0.81	-0.81	-0.81	-0.81	-0.81	-0.81	0.00	0.88	0.88	0.88	0.88	0.88
Resnet50 TL	-1.00	-1.00	-1.00	-1.00	-1.00	-1.00	-1.00	-1.00	0.00	0.00	0.00	0.00	1.00	0.00	0.00
Resnet50 SVM	-0.83	-0.83	-0.83	-0.83	-0.83	-0.83	-0.83	-0.83	-0.83	0.00	0.88	0.00	0.88	0.88	0.00
InceptionV3 TL	0.00	-0.86	-0.86	-0.86	-0.86	-0.86	-0.86	-0.86	-0.86	0.00	0.00	0.00	0.00	0.00	0.00
InceptionV3 SVM	-0.81	-0.81	-0.81	-0.81	-0.81	-0.81	-0.81	0.00	0.00	-0.81	1.00	1.00	1.00	1.00	0.00
Average	-0.70	-0.61	-0.81	-0.70	-0.81	-0.81	-0.82	-0.64	-0.27	-0.14	0.22	0.11	0.43	0.29	0.14
Final Class	C	C	C	C	C	C	C	C	C	C	N	N	N	N	N

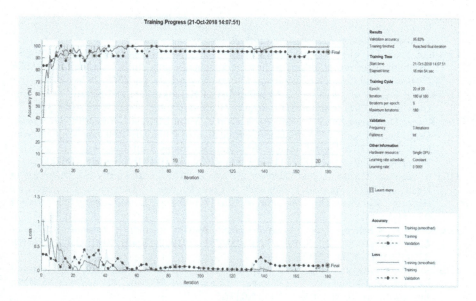

FIGURE 7.7 Sample training graph of VGG 19 model.

7.6 CONCLUSION

Through this paper, we are proposing a robust algorithm to classify the histopathological photomicrographs of the H&E stained tissue taken at the magnification of 10x with respect to the objective lens specification. The proposed algorithm yielded classification result as 100% accurate on both the testing as well as the training data.

This classification algorithm employs seven famous deep learning models, trains them by executing for 20 times on the training data, tests them by executing on the test data, analyzes the results of the models' execution on the testing data, computes the error rates of misclassification, computes the weights on the computed error rates, and finally computes the model rates for the classes. Application of this model rate for the final classification is carried out. In brief, we are fusing the results obtained through different models for the classification process.

REFERENCES

A. Chodorowski, U. Mattsson, and T. Gustavsson. "Oral Lesion Classification Using True Colour Images", *Proceedings of SPIE*, 3661 (1999): 1127–1138. ISBN. 978081943132.

Alex Krizhevsky, Ilya Sutskever, and Geoffrey E. Hinton. "Imagenet classification with deep convolutional neural networks." In *Advances in Neural Information Processing Systems*, pp. 1097–1105. (2012).

B. Kieffer, M. Babaie, S. Kalra, and H. R. Tizhoosh. "Convolutional Neural Networks for Histopathology Image Classification: Training vs. Using Pre-Trained Networks", ArXiv171005726 Cs. (2017).

Christian Szegedy, Wei Liu, Yangqing Jia, Pierre Sermanet, Scott Reed, Dragomir Anguelov, Dumitru Erhan, Vincent Vanhoucke, and Andrew Rabinovich. "Going deeper with convolutions." In *Proceedings of the IEEE Conference on Computer Vision and Pattern Recognition*, pp. 1–9. (2015).

Christian Szegedy, Vincent Vanhoucke, Sergey Ioffe, Jon Shlens, and Zbigniew Wojna. "Rethinking the inception architecture for computer vision." In *Proceedings of the IEEE Conference on Computer Vision and Pattern Recognition*, pp. 2818–2826. (2016).

Dev Kumar Das, et al. "Automated Identification of Keratinization and Keratin Pearl Area from In Situ Oral Histological Images", *Tissue and Cell*, 47(4) (2015): 349–358.

Dev Kumar Das, Bose Surajit, Maiti Asok Kumar, Mitra Bhaskar, Mukherjee Gopeswar, Dutta Pranab Kumar. "Automatic Identification of Clinically Relevant Regions from Oral Tissue Histological Images for Oral Squamous Cell Carcinoma Diagnosis", *Tissue and Cell*, 53 (2018): 111–119. https://doi.org/10.1016/j.tice.2018.06.004

Julia A. Woolgar, Ferlito, A. and Devaney, K.O. et al. "How Trustworthy Is a Diagnosis in Head and Neck Surgical Pathology? A Consideration of Diagnostic Discrepancies (Errors)", *Eur Arch Otorhinolaryngol*, (2011): 643–651. https://doi.org/10.1007/s004 05-011-1526-x

Ken Coelho. "Challenges of the Oral Cancer Burden in India", *Journal of Cancer Epidemiology*. (2012). 701932. 10.1155/2012/701932. Epub 2012 Oct 4. PMID: 23093961; PMCID: PMC3471448.

Kaiming He, Xiangyu Zhang, Shaoqing Ren, and Jian Sun. "Deep Residual Learning for Image Recognition", *Computer Vision and Pattern Recognition*. (2015). https://doi.org/10.48550/arXiv.1512.03385

Karen Simonyan and Andrew Zisserman. "Very deep convolutional networks for large-scale image recognition", *arXiv preprint arXiv:1409.1556*. (2014).

Maisun Mohamed Alzorgani and Hassan Ugail. "Deep Learning Models for Histopathological Images Classification", In *The 2nd Annual Innovative Engineering Research Conference(AIERC)*. (2018). 10.29007/1v9h

M. Muthu Rama Krishnan, et al. "Automated Oral Cancer Identification Using Histopathological Images: A Hybrid Feature Extraction Paradigm", *Micron*, 43(2–3) (2012): 352–364.

M. Sudhakara, V. Reshma, Nawal Khan, and S. R. Amulya. "Uncommon Features In Conventional Oral Squamous Cell Carcinoma", *J Oral Maxillofac Pathol*, 20(2) (2016 May–August): 316–319. 10.4103/0973-029X.185905, PMCID: PMC4989568s.

Navid Noroozi and Ali Zakerolhosseini. "Differential Diagnosis of Squamous Cell Carcinoma In Situ Using Skin Histopathological Images", *Computers in Biology and Medicine*, 70 (2016): 23–39.

Pascal Getreuer. "Chan-Vese Segmentation", *Image Processing On-Line*, 2 (2012): 214–224. 10.5201/ipol.2012.g-cv

Pegah Khosravi, et al. "Deep Convolutional Neural Networks Enable Discrimination of Heterogeneous Digital Pathology Images", *EBioMedicine*, 27 (2018): 317–328. 10.1016/j.ebiom.2017.12.026

R. Chandrakar, R. Raja, R. Miri, R. K. Patra, and U. Sinha. "Computer Succored Vatication of Multi-Object Detection and Histogram Enhancement in Low Vision", *Int. J. of Biometrics. Special Issue: Investigation of Robustness in Image Enhancement and Preprocessing Techniques for Biometrics and Computer Vision Applications*, 3(1) (2020).

Santhosh Kumar Caliaperoumal, R. Vezhavendhan, A. Santha Devy, Priyavendhan, and Uma Devi. "Demonstration and Comparison of Keratin Pearl and Individual Cell Keratin in Oral Squamous Cell Carcinoma using Modified Mallory's Stain and Hematoxylin and Eosin", *International Journal of Current Microbiology and Applied Sciences (IJCMAS)*, 5(7) (2016): 586–591. http://dx.doi.org/10.20546/ijcmas.2016.507.065

Ty Rahman, et al. "Textural Pattern Classification for Oral Squamous Cell Carcinoma", *Journal of Microscopy*, 269(1) (2018): 85–93.

8 Prediction of Stage of Alzheimer's Disease DenseNet Deep Learning Model

Yogesh Kumar Rathore
Shri Shankaracharya Institute of Professional Management and Technology, Raipur, Chhattisgarh, India

Rekh Ram Janghel
Department of Information Technology, National Institute of Technology, Raipur, Chhattisgarh, India

CONTENTS

8.1 Introduction 105
8.2 Literature Survey 107
8.3 Methodology 108
    8.3.1 Deep Learning Techniques 108
    8.3.2 Dataset 108
    8.3.3 Data Preprocessing 108
    8.3.4 Network Architecture 110
    8.3.5 DenseNet 111
8.4 Experiment and Result Discussion 112
    8.4.1 Using Machine Learning 112
    8.4.2 Using Deep Learning 114
8.5 Conclusion 117
References 119

8.1 INTRODUCTION

Alzheimer disease is one of the top ten reasons for death. The disease has not found any curable treatment till date. Thus, it is important to identify the disease at the earliest stage to suggest individuals' proper treatment to improve their lives (Gaugler et al., 2019). As of now, clinical assessments show neuropsychological proof of intellectual disability and clinical history examination are used as the clinical diagnosis of Alzheimer's disease. An accurate diagnosis or identification requires years of experience in the field of neurobiology, resulting into a time-consuming task, and years of training, making it hard for a less experienced clinician to diagnose (Altaf et al., 2018) Subsequently, a computer-aided-diagnosis model would help guide towards the direction of proper diagnosis of the disease. When a patient is affected with Alzheimer disease, it results in two common conditions:

a. There is a decomposition of thick layers of proteins between and outside the nerve cells.
b. There is a presence of neurofibrillary tangles among nerve fibers instead of straight nerve fibers inside the brain.

Thick protein deposition and tangles along the nerve fibers become helpful in diagnosing disease. The ventricles present in the brain grow, whereas the brain tissues and cerebral cortex of the brain shrink (McKhann et al., 2011; Sinha, et al., 2011). These impacts on the brain can be identified easily using MRI scan, which become highly effective in predicting the stage of disease (Tiwari et al., 2021). We present a computer-aided diagnosis (CAD) system in this research where input of binary class is taken by the proposed model, a class of MRI images that contains images of patients who have the disease and class containing of patients who do not the disease. Then, it undergoes some preprocessing and data augmentaion, and we are proposing deep neural network architectures that are pre-trained architectures employed over the ImageNet. Those neural network architectures do not contain fully connected layers or dense. These neural networks perform feature extraction from the MRI based on the principle of transfer learning and forward the results to the a support vector machine classifier for identification of Alzheimer's Disease. In the end, the DensNet deep learning model applies for the training and testing of images of the Alzheimer dataset (Figure 8.1).

Alzheimer disease influences everyone differently. Its effect varies from person to person. The effect of disease mostly depends on the lifestyle and day-to-day work and health of a particular patient. The disease can affect a person in three ways. First, it can affect person so slowly that it difficult to recognize. Sometimes, it affects the person quickly, and sometimes it becomes worse very fast.

Progression of Alzheimer's Disease

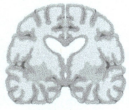
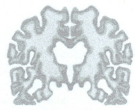

Healthy Brain   **Mild Alzheimer's Disease**   **Severe Alzheimer's Disease**

FIGURE 8.1 Progression of Alzheimer's disease.

In the further sections, we provided a literature survey that comprises work of various authors who have worked on Alzheimer's detection using the Alzheimer Disease Neuroimaging Initiative (ADNI) Dataset, and different methods applied for the same. Further, we discuss the methodology used to detect different stages of Alzheimer's disease and show a detailed explanation of Alzheimer's disease, preprocessing of data, and then finally the deep learning models, including DenseNet169 and DenseNet201, following up with the result and conclusion.

8.2 LITERATURE SURVEY

Tong et al. presented a machine-learning-based approach to detecting Alzheimer disease at an early. They applied one-to-one comparison between the two. Then, in the final stage, all the classes are combined for a final decision (Tong et al., 2016; Raja et al., 2021). M. Prince et al. presented a report in 2013 in which they explained the severity, stages, and challenges of Alzheimer disease. Similarly, many other authors presented the use of machine-learning-based approaches for Alzheimer disease prediction, and one author presented a mixed kernel method with an SVM classifier to categorize Alzheimer disease (Arvesen et al., 2015; Pandey et al., 2022). The mixed kernel is made with the aid of using a fusion of functions from the above-stated three modalities. The present-day studies in PC imaginative and prescient and device studying are prompted with the aid of using neural networks and deep studying. Deep learning is an illustration studying method that permits a device to analyze extraordinary representations from raw information. The purpose at the back of the recognition is the hierarchical and layered shape of the community. Convolution neural networks (CNN) are stimulated with the aid of using the human visible gadget and analyzing the features via a compositional hierarchy of objects, beginning with easy edges and shifting toward greater complicated forms. An alter net combination of convolution and pooling layers was used to extract the features from the input image and reduce the feature with the aid of downsizing the function map (Islam and Zhang, 2017a) after each stage, respectively. Deep studying has additionally inspired neuroscience researchers, and they commenced discovering their answers for issues related to neuro-imaging. Lui et al. (Islam and Zhang, 2017b) used a zero-covering approach for fusing information from special modalities, after which they used a skilled stacked autoencoder (SAE) community for the category. In any other paintings, (Basaia et al., 2018; Raja et al., 2018) used the concept of the usage of multi-segment functions accompanied with the aid of using SAE and a linear in addition to a softmax classifier. They took MMSE as a low degree function and excessive degree function included multimodal neuroimaging records. Shi et al. (Spasov et al., 2019) applied multimodal stacked deep polynomial networks (SDPNs) for the category. Two separate SDPN learned functions from MRI and PET information and the outputs have been then fused and handed to a very last level SDPN. They finished approximately 97% accuracy for diagnosing Alzheimer disease. However, outcomes of those researches have been as accurate as outcomes from the usage of picture processing strategies for the multi-class category. An appropriate development in category accuracy is visible inside the work by Gupta et al. (Anwar et al., 2018), where they first learned bases from natural images and MRI scans in an SAE community, after which they convolved those bases with MRI information to analyze features while training. They acquired 85% accuracy, however, using a one-vs-all approach for three classes. Payan et al. (K. A. N. Gunawardena et al., 2017) used a comparable method; they finished three-D convolution of pre-trained bases with MRI information and handed the function maps to absolutely linked networks. They finished 89.5% accuracy for classifying AD, MCI, and normal in a setup. Hosseini et al. (T. Song et al., 2019) increased the idea of the usage of pre-trained bases and applied three-D convolutional car encoders (CAE) with three special scales to analyze feature bases for the three convolutional layers of CNN architecture. Multiple absolutely linked layers have been used on the pinnacle of convolutional layers for class evaluations. However, Sarraf et al. (Kruthika and Maheshappa, 2019) have been the first to put into effect the diagnosing pipeline with natural CNN and with no pre-training. They trained their model for MRI, in addition to resting-state practical MRI information, one after the other in the usage of LeNet and GoogLeNet fashions and achieved great outcomes for the binary category.

8.3 METHODOLOGY

8.3.1 Deep Learning Techniques

Figure 8.2 shows the proper flow of data stepwise. At first stage, we acquire the image from the ADNI dataset. Then, the task of pre-processing is applied to convert the MRI image into a 2D image. At the third stage, images are divided into different training and testing ratios and fed into the model for further processing of classification. The details of each step are explained below.

8.3.2 Dataset

To perform our experiments and validate our proposed model, we have used the ADNI public dataset (Vaithinathan and Parthiban, 2019). The dataset contains MRI images scanned from a 3 T scanner and comprises four classes, as given in Table 8.1. The images are captured at different time spans, to evaluate the growth of Alzheimer disease year to year. Here, NC represents the images of Normal Character; MCI is the case when the effect of Alzheimer disease is Mild; then, LMCI is when Alzheimer disease is at stage 2, and finally, AD shows images of patients who have confirmed cases of Alzheimer disease. Total training data included 4098 images, and validated data consisted of 254 images, which were tested using 2225 images.

8.3.3 Data Preprocessing

MRI scans are provided in the form of 3D Nifti volumes. At first, skull stripping and gray matter (GM) segmentation is carried out on axial scans through spatial normalization, bias correction, and modulation using the SPM-8* tool.

First, we have converted all the input images into JPEGs using the Nibabel package of Python language. As shown in Figure 8.3, images are very similar and require highly discriminative features for correct diagnosis.

Prediction of Stage of Alzheimer's Disease

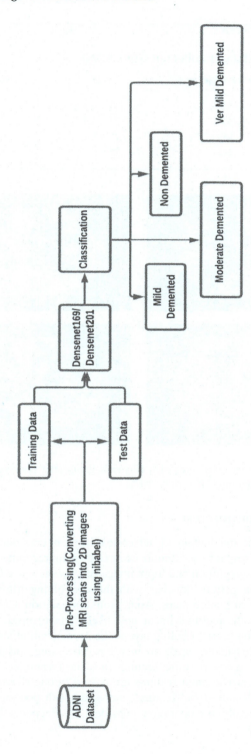

FIGURE 8.2 Flow-chart of proposed methodology.

TABLE 8.1
Details of the MRI dataset collected from ADNI

Classes	No. of Subjects	Age	No. of MRI Volumes
AD	73	74 ± 8.3	717
LMCI	22	73 ± 7.4	52
MCI	149	75 ± 8.2	1792
NC	245	75 ± 3.7	2560

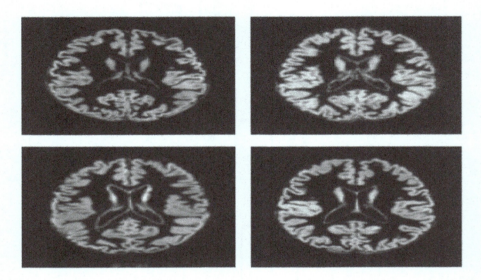

FIGURE 8.3 Gray matter images for each diagnosis group after pre-processing (Top left: AD, right: LMCI, bottom left: MCI, right: NC).

8.3.4 Network Architecture

As mentioned earlier, the human visual cortex inspires convolution neural networks. Each neuron responds to stimuli occurring in its receptive field. This operation is similar to convolution and input image patch works as its small receptive field. When the input pattern matches with the convolving filter, a response is produced in the form of feature maps. Apart from convolution, CNN also includes other layers, such as the rectified linear unit (ReLU) layer used as an activation function for succeeding layers, pooling layers, the normalization layer to normalize the output features in smaller scale to increase processing speed, and the fully connected (FC) layer to covert output features in scalar form to easily connect with the classification layer. The input features are passed through various convolution layers to extract local and global features connected with pooling and Relu activation function. Convolution operation considers local connectivity, parameter sharing, and shift invariance, hence becoming more powerful than traditional feature extraction methods. Finally, features are supplied to fully connected neurons.

Highly complex CNN architectures have been developed so far, including AlexNet (R. Baik, 2019), ZFNet (H. Karasawa et al., 2018), VGG-Net, GoogleNet, and DennseNet from ImageNet challenge. These models are successfully being used in numerous applications like object recognition, detection, segmentation, and more. The proposed framework is also experimented upon two state-of-the-art models, namely DenseNet, the details of which are as follows.

8.3.5 DenseNet

Dense convolutional network (DenseNet) was developed by Kaiming et al. (O. Ronneberger et al., 2015) and was the winner of ILSVRC-2015. The architecture included a central concept of shortcut or skip connections, which are added as a bypass to convolutional layers of the regular feed-forward network; making the block a residual block as shown in Figure 8.4. The main working principle of the DenseNet model is that it not only captures the features from F(x) i.e. immediate previous layer but also captures the features from the x input layer (P. Kochunov et al., 2002). For normal feed-forward networks, the prediction accuracy decreases as the depth of the network increases. Many factors are responsible for such results, including the vanishing gradient problem, saturation, size of training data, and overfitting. Residual learning allows network depth to become as deep as more than a thousand layers. During backward pass, skip connections make the flow of the gradient easy and solve vanishing gradient problems. A ResNet (Liu et al., 2019) with 152 layers is eight times deeper than a VGG (A. Krizhevsky et al., 2012) network but still has lower computational complexity.

Layers	Output Size	DenseNet-121		DenseNet-169		DenseNet-201		DenseNet-264	
Convolution	112 × 112	7 × 7 conv, stride 2							
Pooling	56 × 56	3 × 3 max pool, stride 2							
Dense Block (1)	56 × 56	[1 × 1 conv; 3 × 3 conv]	× 6	[1 × 1 conv; 3 × 3 conv]	× 6	[1 × 1 conv; 3 × 3 conv]	× 6	[1 × 1 conv; 3 × 3 conv]	× 6
Transition Layer (1)	56 × 56	1 × 1 conv							
	28 × 28	2 × 2 average pool, stride 2							
Dense Block (2)	28 × 28	[1 × 1 conv; 3 × 3 conv]	× 12	[1 × 1 conv; 3 × 3 conv]	× 12	[1 × 1 conv; 3 × 3 conv]	× 12	[1 × 1 conv; 3 × 3 conv]	× 12
Transition Layer (2)	28 × 28	1 × 1 conv							
	14 × 14	2 × 2 average pool, stride 2							
Dense Block (3)	14 × 14	[1 × 1 conv; 3 × 3 conv]	× 24	[1 × 1 conv; 3 × 3 conv]	× 32	[1 × 1 conv; 3 × 3 conv]	× 48	[1 × 1 conv; 3 × 3 conv]	× 64
Transition Layer (3)	14 × 14	1 × 1 conv							
	7 × 7	2 × 2 average pool, stride 2							
Dense Block (4)	7 × 7	[1 × 1 conv; 3 × 3 conv]	× 16	[1 × 1 conv; 3 × 3 conv]	× 32	[1 × 1 conv; 3 × 3 conv]	× 32	[1 × 1 conv; 3 × 3 conv]	× 48
Classification Layer	1 × 1	7 × 7 global average pool							
		1000D fully-connected, softmax							

FIGURE 8.4 DenseNet architectures.

In the last few years, research has shown that, if we go deeper in the convolution neural network, the accuracy of classification will increase. Here, we used the DenseNet model (Simonyan and Zisserman, 2014), in which all the layers are connected with the previous and nest layer in a feed-forward manner. Other CNN models with N number of layers (Hon and Khan, 2017) have N number of connections in a one-to-one manner, whereas the DenseNet model has L(L + 1)/2 connections among layers. That means that each layer is connected with all the previous layers. In this way, it is able to extract more effective features. It has various advantages, like they lighten the vanishing gradient issue, reinforce characteristic propagation, inspire characteristic reuse, and drastically reduce the quantity of parameters. We compare our proposed structure on four noticeably competitive item reputation benchmark tasks: CIFAR-10 (A. Farooq et al., 2017), CIFAR-100 (S. Korolev et al., 2017), SVHN (Liu et al., 2017), and ImageNet (M. Boccardi et al., 2015). DenseNets (Chandrakar et al., 2021; Nawaz et al., 2021; Tiwari et al., 2020) acquired full-size enhancements over the state-of-the-art (Noor et al., 2020; Raja et al., 2018) on a maximum of them, while requiring much less memory and computation to acquire excessive performance.

8.4 EXPERIMENT AND RESULT DISCUSSION

8.4.1 Using Machine Learning

Table 8.2 comprises the accuracy comparison of the proposed method. The table contains precision, recall, F1 score, and accuracy of different machine learning classifiers applied on a given dataset. Here, the highest accuracy is recorded for the random forest classifier. The graphical representation of Table 8.2 is shown in Figure 8.5 below.

From Table 8.3, The best accuracy found for n-estimator = 200, criterion = gini, max-depth = (Islam and Zhang, 2017b), and max-feature = auto is 83.93%.

TABLE 8.2
Result comparison of different classifiers

Sl. No.	Method	Precision	Recall	F-1 Score	Accuracy (in %)
1.	Random Forest	86	86	86	85.71
2.	Extra-Trees Classifier	79	78	79	79.46
3.	Decision Tree Classifier	78	78	78	77.67
4.	XgBoost	84	83	84	83.92
5.	Adaboost Classifier	80.5	80	80	80.35
6.	K-Nearest Neighbor Classifier	69	67	67	66.96
7.	Logistic Regression	75	74	74	74.10
8.	NuSVC	79	79	79	79.46
9.	Linear SVC	78	75	75	75
10.	Gradient Boosting Classifier	85	84	85	84.82
11.	Multilayer Neural Network	80	68	74	74.17

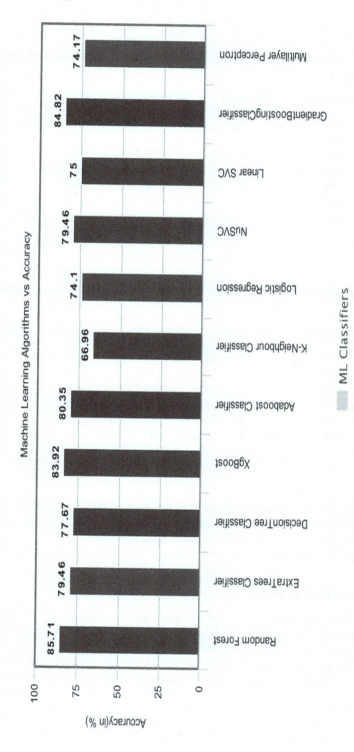

FIGURE 8.5 Machine learning algorithms with their respective accuracy.

TABLE 8.3
Accuracy of random forest classifier while keeping n-estimator and max feature constant

Sl. No.	n-Estimator	Max-depth	Max-feature	Criterion	Accuracy
1.	200	4	auto	gini	78.57
2.	200	5	auto	gini	78.57
3.	200	6	auto	gini	81.25
4.	200	7	auto	gini	83.03
5.	**200**	**8**	**auto**	**gini**	**83.93**
6.	200	9	auto	gini	83.04

TABLE 8.4
Accuracy of random forest classifier keeping max-depth and max feature constant

Sl. No.	n-Estimator	Max-depth	Max-feature	Criterion	Accuracy
1.	20	8	auto	gini	83.03
2.	30	8	auto	gini	84.82
3.	50	8	auto	gini	83.93
4.	70	8	auto	gini	83.93
5.	75	8	auto	gini	82.14
6.	**80**	**8**	**auto**	**gini**	**85.71**
7.	90	8	auto	gini	83.03

From Table 8.4, we can see the maximum accuracy of 85.71%, which is for n = 80. The best accuracy found for n-estimator = 80, criterion = gini, max-depth = (Islam and Zhang, 2017a), and max-feature = auto is 85.71%.

From Table 8.5, we can see the best accuracy is 85.71% at test size 0.3.

8.4.2 Using Deep Learning

From Table 8.6, we observed that the optimal value of dropout is 0.5, which is giving an accuracy of 89.726% in DenseNet169 and 85.689% in DenseNet201, respectively. One line is 0.5 for dropout maximum accuracy.

From Table 8.7, we conclude that the best accuracy for both the models is 93.045% for DenseNet169 and 92.218% for DenseNet201, respectively, and better

TABLE 8.5
Random forest classifier output putting all the parameters constant and varying test and testing size

Sl. No.	Test Size	Accuracy (in %)
1.	0.1	81.57
2.	0.2	78.66
3.	**0.3**	**85.71**
4.	0.4	81.33
5.	0.5	77.00
6.	0.6	81.25
7.	0.7	80.15
8.	0.8	79.93

TABLE 8.6
Keeping activation function fixed and varying dropout in to get optimal value of dropout with 10 epochs

Sl. No.	Activation Function	Dropout	DenseNet169 (accuracy in %)	DenseNet201 (accuracy in %)
1.	Rectified Linear Unit (ReLU)	0.1	85.012	83.034
2.	Rectified Linear Unit (ReLU)	0.2	86.501	84.332
3.	Rectified Linear Unit (ReLU)	0.3	87.502	82.801
4.	Rectified Linear Unit (ReLU)	0.4	87.923	83.270
5.	**Rectified Linear Unit (ReLU)**	**0.5**	**89.726**	**85.689**
6.	Rectified Linear Unit (ReLU)	0.6	87.338	84.405
7.	Rectified Linear Unit (ReLU)	0.7	87.391	84.324

accuracy of 93.045% is obtained with DenseNet169 with activation function as rectified linear unit (ReLU) and dropout as 0.5.

Figure 8.6 and Figure 8.7 show the performance of the model with respect to each epoch.

TABLE 8.7
Keeping dropout fixed and varying activation function to know the optimal activation function for both DenseNet169 and DenseNet201 with 25 epochs

Sl. No.	Activation Function	Dropout	DenseNet169 (accuracy in %)	DenseNet201 (accuracy in %)
1.	Sigmoid	0.5	79.026	74.318
2.	Hyperbolic Tangent Function (Tanh)	0.5	81.324	80.419
3.	**Rectified Linear Unit (ReLU)**	**0.5**	**93.045**	**94.079**
4.	Exponential Linear Units (ELUs)	0.5	89.082	87.912

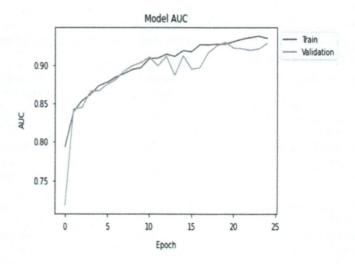

FIGURE 8.6 Accuracy vs epoch.

From the accuracy plot shown in Figure 8.6, we can see that the model accuracy is still increasing after 25 epochs, so we can try for some more epochs. At the same time, from the loss curve shown in Figure 8.7, we can see that training and validation loss both are constant now, so there are no drastic changes expected to be observed for further training. So, as of now, finally we can decide to go with this model.

From Table 8.8, we observed that our proposed method with DenseNet169 and DenseNet201 gives the best accuracy among other proposed models introduced in different research papers.

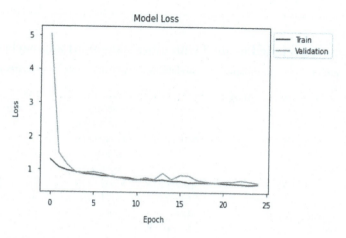

FIGURE 8.7 Loss vs epoch.

From Table 8.9, we can observed that the DenseNet model performs well compared to other models.

8.5 CONCLUSION

Alzheimer disease is one of the most common diseases found in senior citizens worldwide. Early and accurate diagnosis of this disease is important. Here, we used the ADNI dataset, and at the first stage, we applied 11 machine learning algorithms named random forest classification, extra trees classification, decision trees classification, XgBoost classification, Adaboost classification, K-nearest neighbor classification, logistic regression classification, NuSVC (Nu-support vector classifier) classification, linear SVC (support vector classifier) classification, gradient boosting classification, and multi-layer perceptron neural network classification with accuracy of 85.71%, 79.46%, 77.67%, 83.92%, 80.35%, 66.96%, 74.10%, 79.46%, 75%, 84.82%, and 74.17%, respectively. The highest accuracy of 85.71% was recorded for random forest. So, dementia and non-dementia random forest classifiers were proven to be the best approach in our study. In this study, at the second stage, we used a convolution neural network deep learning model to check further possibilities in better classification accuracy. Networks were trained and tested using deep DenseNet121, DenseNet169, and DenseNet201 models. Results from both the models outperformed all the other methods of literature in regard to multiclass classification and an accuracy of 91.835%, 93.045%, and 92.218% was obtained with respect to DenseNet121, DenseNet169, and DenseNet201. As compared to most of the previous works, pre-trained feature learning is not required anymore, and still the network can accurately predict the classes. Class-specific performance gain is also accomplished, improving performance for all classes. The best accuracy of 91.835%, 93.045%, and 92.218% was obtained with respect to DenseNet121, DenseNet169, and DenseMet201 with respect to other models we applied.

TABLE 8.8
Comparison of proposed method with other state-of-art approaches

Sl. No.	Author	Technique	Modalities	Classification	Accuracy (%)
1.		SAE-Zeromask	MRI+PET	4 way (AD/cMCI/ncMCI/NC)	53.8
2.		SDPN	MRI+PET	4 way (AD/cMCI/ncMCI/NC)	57
3.		MPFR	MRI+PET+Clinical	3 way (AD/MCI/NC)	59.19
4.	Tong et al. 2016	NGF	MRI+PET+CSF	3 way	60.2
5.		Combined biomarkers	MRI	3 way	62.7
6.		DW-S2 MTL	MRI+PET+CSF	3 way	62.93
7.		SAE	MRI	3 way	85 with natural image bases, 78.2 with MRI bases
8.		SAE_CNN	MRI	3 way	89.4 with 3D convolution 85.53 with 2D convolution
9.		Quantification of MRI deformation	MRI	3 way	91.74% in differentiating progressive MCI from healthy elderly adults
10.	Proposed Method	DenseNet121	MRI	4 way	91.835
11.	Proposed Method	DenseNet169	MRI	4 way	93.045
12.	Proposed Method	DenseNet201	MRI	4 way	94.079

TABLE 8.9
Comparison with other techniques on the basis of SPE, SEN, and PPV

Author	SPE	SEN	PPV
S. Lui et al., 2007	54.31	–	61.29
Sørensen et al. 2008	–	79	–
DenseNet 121		**64.04**	**72.19**
DenseNet 169	–	**72.31**	**78.56**
DenseNet 201	–	**74.55**	**80.36**

REFERENCES

T. Altaf, S.M. Anwar, N. Gul, M.N. Majeed and M. Majid, "Multi-class Alzheimer's disease classification using image and clinical features", *Biomedical Signal Processing and Control*, vol. 43, pp. 64–74, 2018.

Anwar, S. M., Majid, M., Qayyum, A., Awais, M., Alnowami, M., & Khan, M. K. (2018). Medical image analysis using convolutional neural networks: a review. *Journal of medical systems*, vol. 42, no. 11, pp. 1–13.

E. Arvesen, Automatic classification of Alzheimer's disease from structural MRI (Master's thesis), 2015.

S. Basaia, F. Agosta, L. Wagner, E. Canu, G. Magnani, R. Santangelo, … & Alzheimer's Disease Neuroimaging Initiative. "Automated classification of Alzheimer's disease and mild cognitive impairment using a single MRI and deep neural networks", *NeuroImage: Clinical*, vol. 21, pp. 101645, 2019.

M. Boccardi, M. Bocchetta, F.C. Morency, D.L. Collins, M. Nishikawa, R. Ganzola, M.J. Grothe, D. Wolf, A. Redolfi and M. Pievani, "Training labels for hippocampal segmentation based on the EADC-ADNI harmonized hippocampal protocol", *Alzheimer's Dementia*, vol. 11, pp. 175–183, 2015.

R. Chandrakar, R. Raja and R. Miri, "Animal detection based on deep convolutional neural networks with genetic segmentation", *Multimed Tools and Applications*, vol. 73, no. 2, pp. 1–14, 2021. 10.1007/s11042-021-11290-4

R. Chandrakar, R. Raja, R. Miri, U. Sinha, A.K. Kushwaha and H. Raja, "Enhanced the moving object detection and object tracking for traffic surveillance using RBF-FDLNN and CBF algorithm", *Expert Systems with Applications*, vol. 191, p. 116306, 2021. ISSN: 0957-4174. 10.1016/j.eswa.2021.116306.

A. Farooq, S. Anwar, M. Awais and S. Rehman, "A deep CNN based multi-class classification of Alzheimer's disease using MRI", 2017 IEEE International Conference on Imaging Systems and Techniques (IST), pp. 1–6, 2017.

J. Gaugler, B. James, T. Johnson, A. Marin and J. Weuve, "2019 Alzheimer's disease facts and figures", *Alzheimers & Dementia*, vol. 15, no. 3, pp. 321–387, 2019.

K.A.N.N.P. Gunawardena, R.N. Rajapakse and N.D. Kodikara, "Applying convolutional neural networks for pre-detection of Alzheimer's disease from structural MRI data", 2017 24th International Conference on Mechatronics and Machine Vision in Practice (M2VIP), pp. 1–7, 2017.

M. Hon and N.M. Khan, "Towards Alzheimer's disease classification through transfer learning", 2017 IEEE International Conference on Bioinformatics and Biomedicine (BIBM), pp. 1166–1169, 2017.

J. Islam and Y. Zhang, "A novel deep learning based multi-class classification method for Alzheimer's disease detection using brain MRI data", International Conference on Brain Informatics, pp. 213–222, 2017a.

J. Islam and Y. Zhang, An ensemble of deep convolutional neural networks for Alzheimer's disease detection and classification *arXiv preprint arXiv:1712.01275*, 2017b.

H. Karasawa, C.-L. Liu and H. Ohwada, "Deep 3d convolutional neural network architectures for Alzheimer's disease diagnosis", Asian Conference on Intelligent Information and Database Systems, pp. 287–296, 2018.

P. Kochunov et al., "An optimized individual target brain in the Talairach coordinate system", *Neuroimage*, vol. 17, no. 2, pp. 922–927, 2002.

S. Korolev, A. Safiullin, M. Belyaev and Y. Dodonova, "Residual and plain convolutional neural networks for 3D brain MRI classification", IEEE International Symposium on Biomedical Imaging, 2017.

A. Krizhevsky, I. Sutskever and G.E. Hinton, "Imagenet classification with deep convolutional neural networks", *Advances in Neural Information Processing Systems*, vol. 25, pp. 1097–1105, 2012.

K. Kruthika and H. Maheshappa, "Multistage classifier-based approach for Alzheimer's disease prediction and retrieval", *Informatics in Medicine Unlocked*, vol. 14, pp. 34–42, 2019.

S. Li, et al., "Hippocampal shape analysis of Alzheimer disease based on machine learning methods," *American Journal of Neuroradiology*, vol. 28, no. 7, pp. 1339–1345, 2007. 10.3174/ajnr.A0620

S. Liu, S. Liu, W. Cai, S. Pujol, R. Kikinis and D. Feng, "Early diagnosis of Alzheimer's disease with deep learning", IEEE International Symposium on Biomedical Imaging, 29 April–2 May, IEEE, Beijing, China, 2017.

X. Liu, Q. Xu and N. Wang, "A survey on deep neural network-based image captioning", *The Visual Computer*, vol. 35, no. 3, pp. 445–470, 2019.

G.M. McKhann, et al., "The diagnosis of dementia due to Alzheimer's disease: Recommendations from the National Institute on Aging-Alzheimer's Association workgroups on diagnostic guidelines for Alzheimer's disease", *Alzheimer's & Dementia*, vol. 7, no. 3, pp. 263–269, 2011.

S.G. Mueller, et al., "The Alzheimer's disease neuroimaging initiative", *Neuroimaging Clinics*, vol. 15, no. 4, pp. 869–877, 2005.

H. Nawaz, M. Maqsood, S. Afzal, F. Aadil, I. Mehmood and S. Rho, "A deep feature-based real-time system for Alzheimer disease stage detection", *Multimed Tools and Applications*, vol. 80, no. 28, pp. 35789–35807, 2021.

M.B.T. Noor, N.Z. Zenia, M.S. Kaiser, Mamun S. A. and Mahmud M., "Application of deep learning in detecting neurological disorders from magnetic resonance images: A survey on the detection of Alzheimer's disease, Parkinson's Disease and Schizophrenia", *Brain informatics*, vol. 7, no. 1, pp. 1–21. 2020.

S. Pandey, R. Miri and G.S. Sinha, "AFD filter and E2N2 classifier for improving visualization of crop image and crop classification in remote sensing image", *International Journal of Remote Sensing*, vol. 43, no. 1, pp. 1–26, 2022. 10.1080/01431161.2021.2000062

R. Raja, S. Kumar, S. Choudhary and H. Dalmia, "An Effective Contour Detection based Image Retrieval using Multi-Fusion Method and Neural Network", *Submitted to Wireless Personal Communication, PREPRINT Version-2 available at Research Square*, 2021, DOI: 10.21203/rs.3.rs-458104/v1

R. Raja, T.S. Sinha and R.P. Dubey, "Recognition of human-face from side-view using progressive switching pattern and soft-computing technique", *Association for the Advancement of Modelling and Simulation Techniques in Enterprises. Advance B*, vol. 58, no. 1, pp. 14–34, 2018. ISSN: 1240-4543.

R. Raja, T.S. Sinha, R.K. Patra and S. Tiwari, "Physiological trait based biometrical authentication of human-face using LGXP and ANN techniques (2018)", *International Journal of Information and Computer Security*, vol. 10, no. 2/3, pp. 303–320, 2018.

O. Ronneberger, P. Fischer and T. Brox, "U-net: Convolutional networks for biomedical image segmentation", International Conference on Medical Image Computing and Computer-Assisted Intervention, pp. 234–241, 2015.

K. Simonyan and A. Zisserman, Very deep convolutional networks for large-scale image recognition, arXiv preprint arXiv:1409.1556., 2014.

T.S. Sinha, R.K. Patra and R. Raja, "A comprehensive analysis of human gait for abnormal foot recognition using Neuro-Genetic approach", *International Journal of Tomography and Statistics*, vol. 16, no. W11, pp. 56 –73, 2011. ISSN: 2319-3339. http://ceser.res.in/ceserp/index.php/ijts

T. Song, et al., "Graph convolutional neural networks for Alzheimer's disease classification", 2019 IEEE 16th International Symposium on Biomedical Imaging (ISBI 2019), pp. 414–417, 2019.

Sørensen, L. V., Waldorff, F. B., & Waldemar, G. (2008). Early counselling and support for patients with mild Alzheimeras disease and their caregivers: A qualitative study on outcome. *Aging and Mental Health*, vol. 12, no. 4, pp. 444–450.

S. Spasov, L. Passamonti, A. Duggento, P. Liò and N. Toschi, "A parameter-efficient deep learning approach to predict conversion from mild cognitive impairment to Alzheimer's disease", *Neuroimage*, vol. 189, pp. 276–287, 2019.

L. Tiwari, Raja, V. Awasthi, R. Miri, G.R. Sinha, Monagi H. Alkinani and Kemal Polat, "Detection of lung nodule and cancer using novel Mask-3 FCM and TWEDLNN algorithms", *Measurement*, vol. 172, p. 108882, 2021. ISSN: 0263-2241. 10.1016/j.measurement.2020.108882

L. Tiwari, R. Raja, V. Sharma, R. Miri, "Fuzzy inference system for efficient lung cancer detection", In: Gupta M., Konar D., Bhattacharyya S., Biswas S. (eds.), *Computer Vision and Machine Intelligence in Medical Image Analysis*, vol. 992, pp. 33–41, Springer, Singapore, 2020. 10.1007/978-981-13-8798-2_4

T. Tong, Q. Gao, R. Guerrero, C. Ledig, L. Chen, D. Rueckert and Alzheimer's Disease Neuroimaging Initiative, "A novel grading biomarker for the prediction of conversion from mild cognitive impairment to Alzheimer's disease", *IEEE Transactions on Biomedical Engineering*, vol. 64, no. 1, pp. 155–165, 2016.

K. Vaithinathan and L. Parthiban, "A novel texture extraction technique with T1 weighted MRI for the classification of Alzheimer's disease", *Journal of Neuroscience Methods*, vol. 318, pp. 84–99, 2019.

9 An Insight of Deep Learning Applications in the Healthcare Industry

Deevesh Chaudhary, Prakash Chandra Sharma, and Akhilesh Kumar Sharma
Department of Information Technology, Manipal University Jaipur, India

Rajesh Tiwari
Department of Computer Science & Engineering, CMR Engineering College, Hyderabad, India

CONTENTS

9.1 Introduction	123
9.2 Drug Discovery	124
9.3 Medical Image and Diagnostics	126
9.4 Clinical Trials	127
9.5 Patient Monitoring and Personalized Treatment	128
9.6 Chatbot Using NLP	130
9.7 Health Insurance and Fraud Detection	130
9.8 Medical Diagnosis	131
9.9 Future Development	133
References	133

9.1 INTRODUCTION

A variety of applications, including computer games, face detection and recognition [1], chatbots using natural language processing (NLP), and self-driving cars, have achieved great success with deep learning techniques [2]. In recent years, artificial intelligence (AI) and machine learning have grown in popularity and acceptability. The situation became much more complicated when the Covid-19 outbreak broke out. We saw a rapid digital change and the adoption of disruptive technology across various industries throughout the crisis. Healthcare was one of the industries that may profit greatly from the use of disruptive technologies. AI, machine learning, and deep learning have all become critical components of the industry. Deep learning has had a tremendous impact in healthcare, allowing the industry to improve patient monitoring and diagnosis. Recurrent neural networks, a type of deep learning architecture, can construct chemical

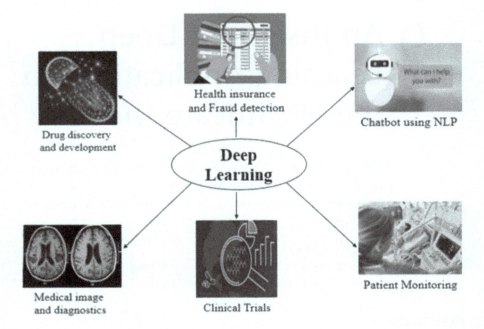

FIGURE 9.1 Application of deep learning technology in the health industry.

structures using pre-programmed principles. With the increasing efficiency of deep learning predictive algorithms and the growing reliance on AI systems for diagnostic assistance, their legitimacy in the healthcare setting is a point of contention. Deep learning is supporting medical practitioners and researchers in uncovering hidden opportunities in data and better serving the healthcare business. Deep learning in healthcare allows clinicians to accurately analyze any ailment and treat it, resulting in improved medical judgments. Figure 9.1 illustrates the application area of deep learning in the health industry. The chapter covers the research work, particularly in these application areas, using deep learning techniques.

9.2 DRUG DISCOVERY

Deep learning plays an important role in discovering medication combinations. Vaccines and drug research were developed with the help of disruptive technologies such as AI, machine learning, and in-depth learning during the epidemic. Due to the complexity of medication research, deep learning can make things faster, cheaper, and easier. Deep learning algorithms can anticipate the pharmacological characteristics, predict drug-target interaction, identify prognostic biomarkers, and create the desired molecule. Deep learning algorithms can easily process genomic, clinical, and population data, and various toolkits can be used to detect patterns between the data. Researchers may now undertake faster molecular modelling and predictive analytics in identifying protein structures using machine learning and deep learning. Figure 9.2 illustrates the drug discovery process in the medical field.

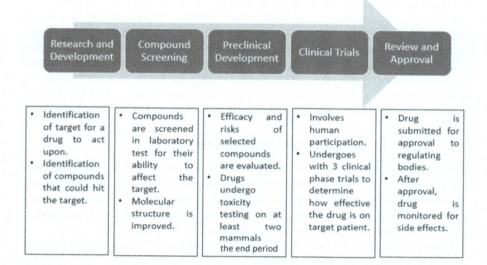

FIGURE 9.2 Steps in drug discovery.

Neves et al. [3] proposed a deep learning model for prediction of anti-plasmodial activity and cytotoxicity of untested compounds. A binary quantitative structure activity relationship model was implemented for selecting the finest model for virtual assessment of huge databases of chemical compounds. Asexual blood stages of Plasmodium falciparum that were both susceptible and multi-drug resistant were used to test the most accurate computer predictions. At low nanomolar doses (EC50 500 nM), two of them, LabMol-149 and LabMol-152, demonstrated significant anti-plasmodial activity in mammalian cells with no cytotoxicity. Using a computer system based on deep learning, the researchers were able to uncover two unique families of possible next-generation antimalarial drugs that satisfied the criteria for antimalarial target candidates.

Mayr et al. [4] won the Tox21 competition and proposed a multitask deep neural network (DNN) model that performed computation on the dataset of 12,000 compounds for 12 high-throughput toxicity tests. Aliper et al. [4] by using transcriptome data from the LINCS project, created DNN models for identifying drug pharmacological characteristics and medication repurposing [5]. DNN models have been proven to predict pharmacological indications with high accuracy using pathway and gene-level data, suggesting that they could be beneficial for drug repurposing.

Lusci et al. [6] proposed an approach that used an newer recurrent neural network (RNN) version known as UGRNN, which maps molecular structure and representation into the vectors of the same length but also thereafter feeds things to a completely interconnected neural network layer to generate replicas. Xu et al. [7] used the similar strategy to simulate the liver damage produced by a medicine. The 475 drugs remaining were used to generate the deep learning models, which were then verified using a 198-drug external dataset. The model had an AUC of 0.955, which was higher than the accuracy of prior DILI models.

Graph convolution models are another important technique for drug discovery. This elementary concept is comparable to the UGRNN technique, which uses neural networks in order to automatically construct a chemical explanation vector using vector data discovered during neural network learning. Morgan's circular fingerprint approach served as inspiration [8]. The neural fingerprint approach described by Duvenaud et al. [9] was one of the earliest attempts to create a graph convolution model. Bjerrum et al. [10] introduced an LSTM RNN that utilises SMILES string as input to create predictive models without needing to generate chemical descriptors.

9.3 MEDICAL IMAGE AND DIAGNOSTICS

Deep learning models can diagnose using medical imagery such as X-rays, MRI scans, and CT scans. In medical photos, the algorithms may detect any risk and indicate irregularities. Deep learning is often employed in cancer detection. Machine learning and deep learning have enabled significant advances in computer vision. It is simpler to treat disorders with a faster diagnosis through medical imaging.

Researchers may examine the phenotypes and behaviors of human and animal hosts, cells, tissues, organs, and other subcellular components using medical imaging. Medical images can be extracted from various sources such as ultrasound images, X-ray images, computed tomography (CT), magnetic resonance imaging (MRI) scan images, positron emission tomography (PET) scans, retinal photography, histology slides images, and dermoscopy. Figure 9.3 shows some example medical images.

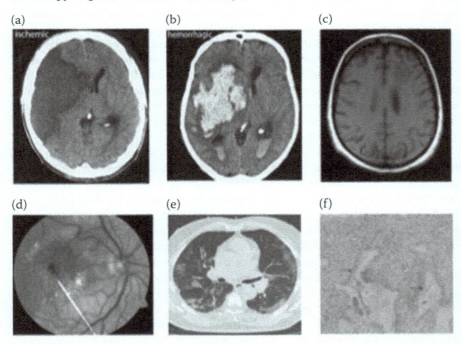

FIGURE 9.3 Medical images (a) Axial CT scan of brain detected with ischemic stroke (b) Hemorrhagic stroke (c) Metastatic disease (d) Retinal Angiomatous disease (e) Chest CT in covid 19 (f) Histology slide of Glioblastoma (brain tumor).

Holger et al. 2017 [11] built a 2-tier cascade approach to run a generation system at high FP levels but with low sensitivity. CAD e-systems were used to generate Volumes of Interest (VOI) using and serving as input for a second tier. A 2D (two-dimensional) or 2.5D view was generated in the second stage using preprocessing techniques such as sampling, scaling, transformations, translations, and random rotations. Deep convolution neural network classifiers are trained using these random views. In the testing phase, the neural network assigns class probabilities to a new set of randomly selected views.

Souza et al. 2019 [12] proposed two CNN architectures used for lung segmentation in order to detect and diagnose pulmonary disorders. The testing of the proposed method was done by using 138 chest X-ray pictures from Montgomery County's Tuberculosis Control Program, with an average sensitivity of 97.54% as the best result.

The Domain Extension Transfer Learning approach for checking COVID-19 symptoms was developed by Basu et al. 2020 [13] using 225 COVID-19 CXR pictures. On a related big chest X-Ray dataset, they used DETL with a pre-trained deep convolution neural network to discriminate between four classes: normal, pneumonia, other disease, and Covid19.

For COVID-19 detection in chest X-Ray pictures, a GAN (Generative Adversarial Network) with deep transfer training was proposed by Loey et al. [14]. Due to the paucity of the COVID-19 dataset, the GAN network was employed to create more X-ray images of chest infection. Masood et al. [15] focused on lung nodule detection/classification and described the multi-dimensional Region-based Fully Convolutional Network (mRFCN), with a classification accuracy of 97.91%. The key task in lung nodule identification seems to be the occurrence of micro-nodules (size < 3 mm) exclusive of compromising the accuracy of model. Wu et al. [16] presented the use of a deep convolution neural network to screen for breast cancer categorization that has been trained and evaluated on over 1,000,000 mammography pictures. Conant et al. [17] proposed an AI method to discover calcified scratches and soft tissue in digital breast tomosynthesis (DBT) images.

9.4 CLINICAL TRIALS

Clinical trials are time-consuming and costly. Machine learning and deep learning techniques can be used to perform predictive analytics to discover possible clinical trial applicants and allow scientists to pool people from many data points and sources. Deep learning will also allow for continuous trial monitoring with minimal human interaction and errors. The high failure rate of trials is due to underprivileged patient group screening and recruitment techniques, and also an incapacity to track patients effectively during studies. Only one drug from every ten that undergo a clinical investigation is approved for commercialization. AI advances may be applied to reimagine critical procedures in the clinical experimental plan, resulting to higher experimental response rates. Various aspects of clinical trials are shown in Figure 9.4.

A Phase I trial has begun for DSP-1181, an AI-created chemical for obsessive-compulsive disorder.[18]. In a Phase II trial of schizophrenia, AiCure, an AI-based mobile application for measuring medication adherence, enhanced compliance by 25% as compared to traditional modified directly observed therapy [19].

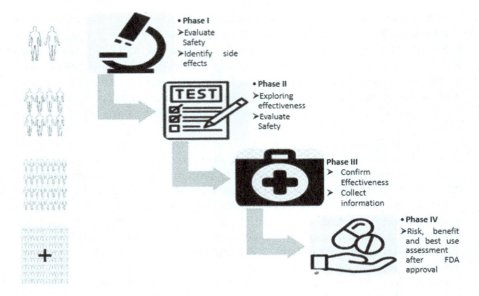

FIGURE 9.4 Phases of clinical trials.

NLP is allegedly used to better match patients to clinical studies, according to the Deep 6 AI [20]. Deep 6's algorithms are trained on thousands of clinical data points taken from patient health records, including symptoms, diagnoses, and test results. The firm claims to be able to carry out the extraction "while disguising protected health information." The clinical data points are then compiled into a clinical profile. Clinical profiles are used to locate and compare groups, cohorts, and individuals who satisfy the user's search criteria. In less than an hour, the Deep 6 AI platform identified 16 patients who met the trial's eligibility criteria.

AI [21] claims to use NLP to assist researchers in handling clinical trial operations more efficiently. Trials are driven by algorithms. To identify risk variables and make recommendations for clinical trial optimization, AI's cloud-based Study Optimizer platform is supposedly trained on "billions of data points from historical clinical trials, medical publications, and real-world sources." Brite Health [22] says that machine learning is being used to improve patient involvement in clinical studies. The algorithm that powers the company's patient app and site dashboard is said to have been trained on millions of clinical data points. The recommendation engine can recognize crucial markers that are linked to patient disengagement from research projects.

9.5 PATIENT MONITORING AND PERSONALIZED TREATMENT

It is easier to assess a patient's health data, medical history, vital symptoms, medical test results, and other information using deep learning models. As a result, healthcare providers are better equipped to comprehend each patient and give them tailored treatment. These game-changing technologies enable the identification of

Deep Learning in Healthcare Sector 129

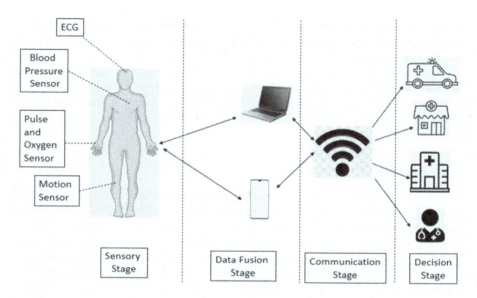

FIGURE 9.5 Architecture of patient monitoring system.

appropriate and multiple treatment alternatives for various individuals. Machine learning models can employ DNNs to forecast upcoming health issues or dangers and deliver appropriate medicines or treatments based on real-time data collected through linked devices. Patient monitoring is illustrated in Figure 9.5.

Machine learning models can be helpful in processing and analyzing both structured and unstructured medical data. It may become challenging to manually classify documents and keep up-to-date health data. As a result, smart health data may be tracked using machine learning and its subset deep learning. With the introduction of telemedicine, wearables, and remote patient monitoring, there is now an abundance of real-time health data, and deep learning can aid in intelligently monitoring patients and predicting hazards.

A smart healthcare IOT technology-based student healthcare monitoring model has been developed by Souri et al. [23] that continually monitors a few factors such as the students' vital signs and detects biological and behavioral changes with these signs. With the help of IOT devices and sensors, data are collected and analyzed using machine learning methods to detect the potential hazards of physiological and behavioral changes in students. Out of multiple machine learning algorithms, the support vector machine attained the best accuracy of 99.1% after assessing the suggested model.

Davoudi et al. [25] presented a model to using IOT and AI technology for granular and autonomous monitoring. This research work is focused on patients in the Intensive Care Unit (ICU) along with a description of psychotic patients and their surroundings. To collect data on patients and their surrounding environment, the researchers employed a camera and a variety of wearable sensors, including light and sound sensitive sensors. They observed various patient-related data collected with the help of sensors and nodes to detect and distinguish the patient's facial expression, body posture, head posture variation, and leg movements, as well as noise pressure levels and light intensity levels of surroundings.

Wong et al. [24] proposed to construct a real-time machine health monitoring system that may examine the supply balancing situation on a three-level structure utilizing machine learning and IOT technology. This system is made up of tiny physical components that can collect and transmit electrical data from a load to a server.

9.6 CHATBOT USING NLP

For classification and identification, deep learning techniques are used in natural language processing (NLP). These two technologies may be applied to recognize and categorize health data, as well as to create chatbots and speech bots. Chatbots play a critical part in today's telemedicine environment. They facilitate and speed up interactions with patients. These chatbots were also used to provide information on COVID-19 and respond to common questions. In many cases, a chatbot serves as a virtual assistant. It is possible to virtualize it. AI is used to replicate conversational abilities and other behaviors.

A study was conducted on how breast cancer patients reply to the chatbot Vik. Chatbot Vik connects with patients through writing messages, responds to any questions or worries, then also reminds patients of their prescriptions. Although human connection and empathy cannot be replaced, 93.95% of the 4737 patients examined recommended Vik to their friends. Patients want tailored, focused assistance and treatment, according to this survey.

Kbot is a customized chatbot with information that was developed for asthma patients for self-management [25]. Nursebot [26] is a chatbot that can assist with self-management of chronic disease. By super positioning, it emulates anatomy (judgement) and shows self-organization, as well as cognitive and emotional responses. It allows for communicative healthcare management. PARRY, a popular chatbot, was created to impersonate a paranoid patient [27]. The popularity of mental health chatbots has grown to the point where companies have dubbed chatbots as the "the future of treatment." A digital chatbot to assist medical staff by reducing the chance of older persons being admitted to the hospital was also investigated, with encouraging results.

9.7 HEALTH INSURANCE AND FRAUD DETECTION

Fraud detection in the healthcare industry has always depended largely on domain experts' knowledge, which is inefficient, expensive, and time intense. Healthcare scams are caught manually by a few personnel who meticulously examine and detect suspicious medical insurance claims, a process that takes a long time and a lot of effort. However, recent advancements in AI and big data methods have improved the efficiency and automation of healthcare fraud detection. There has been a surge of interest in mining healthcare data, mostly for the purpose of detecting and preventing fraud. Deep learning is a powerful tool for detecting insurance fraud and predicting future hazards. Deep learning gives health insurance providers an advantage since the models can predict future trends and behavior, allowing them to recommend smart insurance policies to their consumers.

Srinivasan et al. [28] developed an approach for detecting anomalies in insurance claims data obtained from Medicare data using rule-based data mining, an unsupervised technique. Branting et al. [29] used a supervised technique using labelled dataset, graph theory, and decision tree to analyze health-related data from Medicare facilities. They proposed a technique for assessing the likelihood of healthcare scams based on network algorithms applied to graph theory on open-source information data.

Using CMS data from 2012, Ko et al. [32] focused on the urology sector. By examining heterogeneity among urologists within the service usage and payment, the authors calculated probable savings from uniform service utilization.

The 2013 CMS dataset was used to develop a machine learning model that aimed to see whether medical insurance agents had any unusual conduct when it came to medical insurance claims. It attempts to find out when physicians depart from the standard procedures, such as billing abnormalities, fraud, or lack of knowledge of billing processes. The model was evaluated using various metrics, such as precision and recall calculated by using the five-fold cross-validation technique. The multinomial naive Bayes algorithm is used in the research [30].

9.8 MEDICAL DIAGNOSIS

Practitioners may identify illnesses earlier and more efficiently by studying various kinds of data that affect the health of patients. These types of data include physiological data related to various organs of the human body, external environmental variables, and previous hereditary factors. Clinical data analysis allows us to better understand the underlying processes that underpin illnesses and how risk factors impact their progression. Machine learning enables us to create models that link a variety of factors to an illness. Thankfully, doctors now have access to a wealth of information, ranging from clinical symptoms to biochemical testing and imaging equipment outputs. Figure 9.6 illustrates the use of a machine learning model in medical diagnosis.

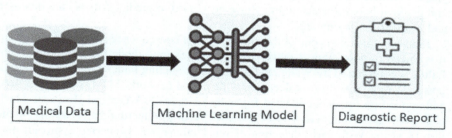

FIGURE 9.6 Medical diagnosis.

It is tough to make decisions in the case of a medical examination. Doctors might make an incorrect diagnosis because of a lack of visual acuity due to distractions, exhaustion, and human visual system limits. Machine learning technologies have been used to help physicians overcome these constraints and make more informed and accurate illness diagnostic judgments. Machine learning approaches have been used to diagnose a variety of disorders. Various research work on machine learning approaches for medical diagnosis of many illnesses are discussed in the following section.

Urinary tract infection (UTI) can be caused by urinary tract inflammation or by some other infection, symptoms of which are similar to each other. Ozkan et al. [31] developed an artificial neural network (ANN) model to improve the diagnostic performance of urinary tract infection. They have collected data of 59 UTI patients and composed a UTI dataset The majorly used classification models, namely decision tree, support vector machine, random forest, and AI, are used to classify between the most common UTIs, i.e. cystistis and urethritis. ANN has achieved the highest classification accuracy of 98.3%.

Pediatric traumatic brain injury (TBI) is very common in children aged less than 15 years. Diagnosing TBI is a challenging task for practitioners. Chong et al. [32] attempted to study the feasibility of using a machine learning algorithm for prediction of TBI, specifically for populations aged less than 16 years. To predict moderate to severe TBI, they developed a multivariable logistic regression model. The logistic regression model uses four variables i.e. road accident, loss of consciousness, vomiting, and symptoms of base skull fracture, whereas machine learning uses another three variables i.e. seizure presence, confusion, and skull fracture. The machine learning model achieved improved accuracy of 98.0% over logistic regression (93.0%) with respect to area under ROC.

Diarrhea is a leading cause of death among people of all ages around the world. Abubakar et al. [33] established a model for predicting diarrhea based on historical data that have been collected from 2013 Nigeria Demographic and Health Survey data, and they developed an ANN model for diarrhea incidence prediction. The study proposed that by using 44 demographic, environmental, and socioeconomic factors, the model achieved an accuracy of 95.63%.

Raita et al. [34] have used demographic and clinical data to develop a deep learning model for prediction of critical care outcomes. The model uses demographic and clinical data extracted from National Hospital and Ambulatory Medical Care Survey (NHAMCS) ED data, from 2007 to 2015 for population aged above 18 years. Four machine learning models, namely random forest, decision tree, DNN, and gradient boosted decision tree, have been used to predict critical care outcome and hospitalization outcome.

Renal dysfunction is one of the most prevalent consequences of heart failure, and it is linked to poor clinical outcomes. A deep learning model for predicting renal failure was proposed by Wang et al. [35]. For predicting fatal complications during hospitalization, a multi-task deep and wide neural network (MT-DWNN) was used and applied on the data collected from the Chinese PLA General Hospital, which had 35,101 heart failure diagnoses and 2478 renal dysfunction diagnoses during the last 18 years. With early detection of renal failure, the suggested approach may reduce the risk of negative effects.

9.9 FUTURE DEVELOPMENT

Machine learning and deep learning methods often require massive datasets for training; however, the human brain can learn from just a few examples. As a consequence, the most prominent difficulties in machine learning algorithms for learning with a small quantity of data have been identified. Matching networks are a deep learning example of using secondary data to improve accuracy of the model by using limited data points [36], which was offered as a one-shot learning version. When the supplementary data were included, the results improved. One-shot learning methods are useful in drug discovery, as chemists frequently work on original targets with insufficient and limited datasets. On cheminformatics datasets, the LSTM approach was used to develop machine learning models with a very limited training set, and results reported are encouraging. A novel sort of deep learning architecture has lately been used by Altae-Tran et al. [37]. A differentiable neural computer (DNC) dramatically enhanced this design. DNCs have been used to solve a variety of issues, including question-and-answer systems, and in graph theory to identify the graph's shortest path, though these additional complex structures have yet to be used in the drug development process.

REFERENCES

[1] A. T. Lopes, E. de Aguiar, A. F. De Souza, and T. Oliveira-Santos, "Facial expression recognition with convolutional neural networks: Coping with few data and the training sample order," *Pattern Recognit*, vol. 61, pp. 610–628, January 2017.

[2] D. Chaudhary, S. Kumar, and V. S. Dhaka, "Estimating crowd size for public place surveillance using deep learning." In: Ahmed K. R. and Hassanien A. E. (eds.), *Deep Learning and Big Data for Intelligent Transportation: Enabling Technologies and Future Trends*, vol. 945, Springer International Publishing, Cham, pp. 175–197. 2021.

[3] B. J. Neves, R. C. Braga, V. M. Alves, M. N. N. Lima, G. C. Cassiano, E. N. Muratov, F. T. M. Costa, and C. H. Andrade, "Deep Learning-driven research for drug discovery: Tackling Malaria," *PLoS Comput. Biol.*, vol. 16, no. 2, p. e1007025, February 2020.

[4] A. Aliper, S. Plis, A. Artemov, A. Ulloa, P. Mamoshina, and A. Zhavoronkov, "Deep learning applications for predicting pharmacological properties of drugs and drug repurposing using transcriptomic data," *Mol. Pharm.*, vol. 13, no. 7, pp. 2524–2530, July 2016.

[5] L. Tiwari, R. Raja, V. Sharma, and R. Miri, "Fuzzy inference system for efficient lung cancer detection." In: Gupta M., Konar D., Bhattacharyya S., Biswas S. (eds.), *Computer Vision and Machine Intelligence in Medical Image Analysis. Advances in Intelligent Systems and Computing*, vol. 992. pp. 33–41. Springer, Singapore. 2020. 10.1007/978-981-13-8798-2_4

[6] R. Chandrakar, R. Raja, R. Miri, U. Sinha, A. K. S. Kushwaha, and H. Raja, "Enhanced the moving object detection and object tracking for traffic surveillance using RBF-FDLNN and CBF algorithm," *Expert Systems with Applications*, Vol. 191, pp. 116306, 2022. ISSN 0957-4174. 10.1016/j.eswa.2021.116306

[7] Y. Xu, Z. Dai, F. Chen, S. Gao, J. Pei, and L. Lai, "Deep learning for drug-induced liver injury," *J. Chem. Inf. Model.*, vol. 55, no. 10, pp. 2085–2093, October 2015.

[8] H. L. Morgan, "The generation of a unique machine description for chemical structures-a technique developed at chemical abstracts service," *J Chem Doc*, vol. 5, no. 2, pp. 107–113, May 1965.

[9] D. Duvenaud, D. Maclaurin, J. Aguilera-Iparraguirre, R. Gómez-Bombarelli, T. Hirzel, A. Aspuru-Guzik, and R. P. Adams, "Convolutional networks on graphs for learning molecular fingerprints," *arXiv*, September 2015.

[10] S. Pandey, R. Miri, G. R. Sinha, and R. Raja,AFD filter and E^2N^2 classifier for improving visualization of crop image and crop classification in remote sensing image," *International Journal of Remote Sensing*, vol. 43, no. 1, pp. 1–26, 2022. 10.1080/01431161.2021.2000062

[11] H. R. Roth, L. Lu, J. Liu, J. Yao, A. Seff, K. Cherry, L. Kim, and R. M. Summers, "Improving computer-aided detection using convolutional neural networks and random view aggregation," *IEEE Trans. Med. Imaging*, vol. 35, no. 5, pp. 1170–1181, 2016.

[12] J. C. Souza, J. O. Bandeira Diniz, J. L. Ferreira, G. L. França da Silva, A. Corrêa Silva, and A. C. de Paiva, "An automatic method for lung segmentation and reconstruction in chest X-ray using deep neural networks," *Comput Methods Programs Biomed*, vol. 177, pp. 285–296, August 2019.

[13] R. Chandrakar, R. Raja, and R. Miri, "Animal detection based on deep convolutional neural networks with genetic segmentation," *Multimed Tools and Applications*, 2021. vol. 73, no. 2, pp. 1–14. 10.1007/s11042-021-11290-4

[14] M. Loey, F. Smarandache, and N. E. M. Khalifa, "Within the lack of chest COVID-19 X-ray dataset: A novel detection model based on GAN and deep transfer learning," *Symmetry*, vol. 12, no. 4, p. 651, April 2020.

[15] L. Tiwari, R. Raja, V. Awasthi, R. Miri, G. R. Sinha, and M. H. Alkinani, Polat, Detection of lung nodule and cancer using novel Mask-3 FCM and TWEDLNN algorithms," *Measurement*, vol. 172, 2021, 108882, ISSN 0263-2241. 10.1016/j.measurement.2020.108882

[16] N. Wu, J. Phang, J. Park, Y. Shen, Z. Huang, M. Zorin, S. Jastrzebski, T. Fevry, J. Katsnelson, E. Kim, S. Wolfson, U. Parikh, S. Gaddam, L. L. Y. Lin, K. Ho, J. D. Weinstein, B. Reig, Y. Gao, H. Toth, K. Pysarenko, A. Lewin, J. Lee, K. Airola, E. Mema, S. Chung, E. Hwang, N. Samreen, S. G. Kim, L. Heacock, L. Moy, K. Cho, and K. J. Geras, "Deep neural networks improve radiologists' performance in breast cancer screening," *IEEE Trans. Med. Imaging*, vol. 39, no. 4, pp. 1184–1194, April 2020.

[17] E. F. Conant, A. Y. Toledano, S. Periaswamy, S. V. Fotin, J. Go, J. E. Boatsman, and J. W. Hoffmeister, "Improving accuracy and efficiency with concurrent use of artificial intelligence for digital breast tomosynthesis," *Radiol. Artif. Intell.*, vol. 1, no. 4, p. e180096, July 2019.

[18] "DSP-1181: Drug created using AI enters clinical trials." [Online]. Available: https://www.europeanpharmaceuticalreview.com/news/112044/dsp-1181-drug-created-using-ai-enters-clinical-trials/. [Accessed: 12th July 2021].

[19] K.-K. Mak and M. R. Pichika, "Artificial intelligence in drug development: Present status and future prospects," *Drug Discov. Today*, vol. 24, no. 3, pp. 773–780, March 2019.

[20] R. Raja, S. Kumar, and M. R. Mahmood, "Color object detection based image retrieval using ROI segmentation with multi-feature method," *Wireless Pers Commun*, vol. 112, pp. 169–192, 2020. 10.1007/s11277-019-07021-6

[21] R. Raja, R. k. Patra, and T. S. Sinha,Extraction of Features from Dummy face for improving Biometrical Authentication of Human, *International Journal of Luminescence and Application*, ISSN:1 2277-6362, vol. 7, no. 3–4, Article 259, pp. 507–512, October–December 2017.

[22] "Brite health care – Medical equipment, CPAP, TENS, Lymphedema compression, traction." [Online]. Available: https://www.britehealthcare.com/. [Accessed: 12th July 2021].

[23] A. Souri, M. Y. Ghafour, A. M. Ahmed, F. Safara, A. Yamini, and M. Hoseyninezhad, "A new machine learning-based healthcare monitoring model for student's condition diagnosis in Internet of Things environment," *Soft comput.*, vol. 24, no. 22, pp. 17111–17121, November 2020.

[24] T. K. Wong, H. K. Mun, S. K. Phang, K. L. Lum, and W. Q. Tan, "Real-time machine health monitoring system using machine learning with IoT technology," *MATEC Web of Conferences*, vol. 335, p. 02005, 2021.

[25] D. Kadariya, R. Venkataramanan, H. Y. Yip, M. Kalra, K. Thirunarayanan, and A. Sheth, "kBot: Knowledge-enabled personalized chatbot for asthma self-management.," *Proc Int Conf Smart Comput SMARTCOMP*, vol. 2019, pp. 138–143, June 2019.

[26] J. P. T. Hernandez, "Network diffusion and technology acceptance of a nurse chatbot for chronic disease self-management support: A theoretical perspective," *J Med Invest*, vol. 66, no. 1.2, pp. 24–30, 2019.

[27] "A chatbot system as a tool to animate a corpus." [Online]. Available: https://1library.net/document/nzwn8mvz-chatbot-tool-animate-corpus.html. [Accessed: 12th July 2021].

[28] U. Srinivasan and B. Arunasalam, "Leveraging big data analytics to reduce healthcare costs," *IT Prof.*, vol. 15, no. 6, pp. 21–28, November 2013.

[29] L. K. Branting, F. Reeder, J. Gold, and T. Champney, "Graph analytics for healthcare fraud risk estimation," in *2016 IEEE/ACM International Conference on Advances in Social Networks Analysis and Mining (ASONAM)*, 2016, pp. 845–851.

[30] R. Chandrakar, R. Raja, R. Miri, R. K. Patra, and U. Sinha, "Computer succored vaticination of multi-object detection and histogram enhancement in low vision," *Int. J. of Biometrics. Special Issue: Investigation of Robustness in Image Enhancement and Preprocessing Techniques for Biometrics and Computer Vision Applications*, vol. 3, no. 1, pp. 1–12, 2021.

[31] I. A. Ozkan, M. Koklu, and I. U. Sert, "Diagnosis of urinary tract infection based on artificial intelligence methods," *Comput Methods Programs Biomed*, vol. 166, pp. 51–59, November 2018.

[32] S. Kumar, R. Raja, and A. Gandham, "Tracking an object using traditional MS (Mean Shift) and CBWH MS (Mean Shift) algorithm with Kalman filter." In: Johri P., Verma J., Paul S. (eds.), *Applications of Machine Learning. Algorithms for Intelligent Systems*. Springer, Singapore. pp. 47–65. 2020. 10.1007/978-981-15-3357-0_4

[33] I. R. Abubakar and S. O. Olatunji, "Computational intelligence-based model for diarrhea prediction using Demographic and Health Survey data," *Soft comput.*, vol. 24, no. 7, pp. 5357–5366, April 2020.

[34] Y. Raita, T. Goto, M. K. Faridi, D. F. M. Brown, C. A. Camargo, and K. Hasegawa, "Emergency department triage prediction of clinical outcomes using machine learning models," *Crit. Care*, vol. 23, no. 1, p. 64, February 2019.

[35] B. Wang, K. He, Y. Bai, Z. Yao, J. Li, W. Dong, Y. Tu, W. Xue, Y. Tian, and Y. Wang, "A multi-task neural network architecture for renal dysfunction prediction in heart failure patients with electronic health records," *IEEE Access*, vol. 7, pp. 178392–178400, 2019.

[36] L. Tiwari, R. Raja, V. Sharma, and R. Miri, "Adaptive neuro fuzzy inference system based fusion of medical image," *International Journal of Research in Electronics and Computer Engineering*, vol. 7, no. 2, pp. 2086–2091, 2020. ISSN: 2393-9028 (PRINT) |ISSN: 2348-2281 (ONLINE).

[37] H. Altae-Tran, B. Ramsundar, A. S. Pappu, and V. Pande, "Low data drug discovery with one-shot learning," *ACS Cent. Sci.*, vol. 3, no. 4, pp. 283–293, April 2017.

10 Expand Patient Care with AWS Cloud for Remote Medical Monitoring

Parul Dubey
Department of Information Technology, Shri Shankaracharya Institute of Professional Management and Technology, Raipur, India

Pushkar Dubey
Department of Management, Pandit Sundarlal Sharma (Open) University, Chhattisgarh, India

CONTENTS

10.1 Introduction 138
10.2 Literature Review 138
10.3 Cloud Healthcare Management 140
    10.3.1 Cloud Computing Models 140
        10.3.1.1 Infrastructure as a Service (IaaS) 140
        10.3.1.2 Platform as a Service (PaaS) 140
        10.3.1.3 Software as a Service (SaaS) 140
    10.3.2 Deployment Models 140
        10.3.2.1 Cloud 141
        10.3.2.2 Hybrid 141
        10.3.2.3 On-Premises 141
    10.3.3 Advantages of Cloud Computation 142
10.4 Amazon Web Services 143
    10.4.1 Compute 143
        10.4.1.1 Amazon Elastic Compute Cloud 143
        10.4.1.2 AWS Elastic Beanstalk 144
    10.4.2 Storage 144
        10.4.2.1 Amazon Elastic Block Store (EBS Volumes) 144
        10.4.2.2 Amazon Elastic File System (Amazon EFS) 144
    10.4.3 Amazon Machine Learning 145
    10.4.4 Big Data Analysis in AWS 145

DOI: 10.1201/9781003217091-10

10.5 Cloud Pricing Strategy 145
10.6 AWS – Healthcare Solutions 147
10.7 Conclusion 147
References 147

10.1 INTRODUCTION

A technical definition of cloud computing in healthcare is "the storage, management, and processing of patient data on remote servers that are accessible over the internet." In contrast to this, setting up a data centre with servers on-site or hosting the files contained on a home computer in the company's office are also viable options.

Hospitals and healthcare providers may now store enormous volumes of data in a secure setting using a network of servers that may be accessed from anywhere, which is made possible by cloud storage, which offers a customizable solution.

After the adoption of the electronic medical records requirement in 2012, healthcare professionals throughout the United States have increasingly relied on cloud-based medical approches to store and preserve patient information (Dhilawala, 2019).

The healthcare industry is one sector where the ageing population has a significant impact. True, many senior citizens have chronic ailments, such as diabetes or high blood pressure, that need continual medical care. However, this is not the case for everyone. As the elderly population continues to grow, hospitals may soon find themselves unable to meet the demands of an ever-growing patient population. Consequently, new approaches to aged care are necessary to enhance and simplify the everyday lives of healthcare workers, while also supporting the elderly in preserving their own health and independence, as well as reducing the drain on the country's financial resources.

Persons with limited mobility, those who reside in remote areas, or those whose health needs require frequent monitoring are the topic of this article, which discusses remote patient monitoring for these groups. It is more accurate to say that we are providing a way for real-time monitoring of delocalized patient health data. Several cutting-edge technologies are used in this discussion, including medical sensor devices, the Internet of Things (IoT), AWS (Amazon Web Services), and cloud computing.

10.2 LITERATURE REVIEW

Cloud computing is a fascinating technology that makes use of software, infrastructure, and the entire computing platform to provide a service. Cloud technology, with the exception of traditional web hosting firms, provides services that are pay-as-you-go in nature. Instead of paying for resources ahead of time, users pay solely for the resources they utilize over time (Chauhan and Kumar, 2013). Patients may find it easier to find and keep track of their status on health records if they use cloud computing (Grogan, 2006).

When it comes to healthcare applications, cellular networks have become more popular (Blake, 2008). Doctor-to-doctor communication is becoming more important in India because of the continuous expansion of mobile phone networks in rural locations that are far from urban areas (Bali and Singh, 2007) Mobile technology is progressively complementing the capabilities of portable devices, including smart phones and PDAs, which have the potential to replace technology-based alternatives while meeting the mobility needs of patients and medical practitioners. With these benefits in mind, it is becoming increasingly feasible to create unique technologies for impoverished territories that might alter the delivery of services and ease the problems caused by healthcare delivery systems (Perera, 2009).

The cloud computing approach is comprised of services rather than applications, as opposed to on-premises computing (Buyya, Yeo, and Venugopal, 2008). A virtualized resource as a service that can be evaluated and paid for is provided by this model. As a result of these benefits, it is becoming easier to generate new ideas. Developing markets with innovative and creative applications have the potential to make a difference. Service delivery and problem resolution methods should be looked for by the healthcare system. The risks to security and privacy have only escalated as a result of this. In the healthcare industry, there is no question that cloud computing may be beneficial. There are several benefits to be achieved by using computer-based solutions: improving patient happiness while lowering overall healthcare costs (Muir, 2011).

Security of patient data continues to be an issue in healthcare systems, but great progress has been achieved in this area as well. Nacha and Pattra, for example, developed a mobile and secure solution that was hosted in the cloud. On the basis of WSN and the cloud, the team in (Maria et al., 2009) presented a comprehensive healthcare system that would also include the deployment of Cipher text Policy-ABE. A DRM (Data Relationship Management) service that safeguards the security of data transfers was discussed in detail in (Hoon and Libor), which presented an outline of how any future cloud-based healthcare system may decrease stress while simultaneously increasing security. Due to this, current trends encourage the use of Blockchain technology to improve security (Mohammed and Maddikunta, 2014). Decentralized databases are combined with encryption in this technology, which has previously shown its usefulness in other areas, such as cryptocurrencies.

AWS is a service provider. One may consider this service to be an exceptional example of genuine cloud computing, since it delivers excellent cloud services while also ensuring the security, integrity, and availability of the data belonging to the clients. AWS offers the resources necessary to develop an application, as well as aid in deploying it across international borders, for a cheap cost to users. As a consequence, businesses have generally concentrated their efforts on a single geographic region at a time, realizing that it was difficult to provide high-quality service to all of their clients. Thanks to AWS, this problem has been rectified, enabling companies throughout the globe to provide a better user experience to their customers. It is an incredible example of actual cloud computing since not only does the service offer cloud computing services, but it also ensures the privacy, accuracy, and availability of your information (Blake, 2008). It is also free.

10.3 CLOUD HEALTHCARE MANAGEMENT

10.3.1 Cloud Computing Models

Three main cloud computing models are discussed here. Each model has a different role in the computation stack.

10.3.1.1 Infrastructure as a Service (IaaS)

The phrase Infrastructure as a Service (IaaS) may be summarized as follows: it is the essential building component of cloud computing, On a subscription basis, cloud computing services provide customers with access to network resources, machines (virtual or dedicated hardware), and computer storage space. It is the most comparable to current information technology (IT) resources. When you choose an IaaS provider, you have the maximum amount of control and flexibility over your IT infrastructure.

10.3.1.2 Platform as a Service (PaaS)

Organizations may devote their resources to building, deploying, and administering their applications instead of to maintaining and managing the underlying infrastructure (typically hardware and operating systems). The ability to focus on other tasks is made easier when you don't have to worry about obtaining the resources you need, planning for capacity, or upgrading and patching your program as often.

10.3.1.3 Software as a Service (SaaS)

You may make use of a completed product that is run and maintained on your behalf by the service provider (SaaS). When most people talk about SaaS, they're referring to software that users interact with. When a SaaS product is offered via the cloud, there is no need to be concerned with the upkeep or administration of the product; instead, think on how you will use a certain piece of software. It is possible to respond to emails via the Internet using a SaaS service, which eliminates the need to administer the infrastructure and operating systems around which the email application is running. Among the types of applications that fall under this category are web-based email services that enable you to secure communications without having to deal with the infrastructure and operating system on which the email system is running.

Figure 10.1 describes the cloud healthcare management service. This diagram explains how cloud takes care of minor things. This can be obtained at less cost, ensuring higher security and availability and ultimately better service and patient care. Cloud manages healthcare in a smooth way. This can be achieved by automation of the system. It shows the management in a diagrammatic view. It shows how it can manage all the facilities related to healthcare under one roof.

10.3.2 Deployment Models

Let's discuss some of the cloud computing deployment models.

AWS Cloud for Remote Medical Monitoring

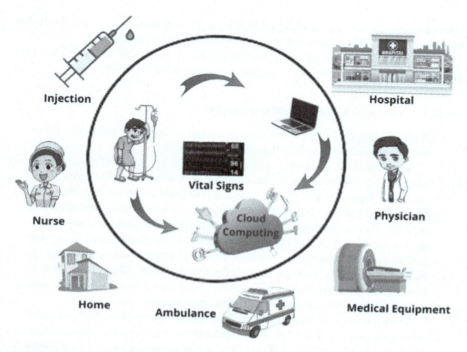

FIGURE 10.1 Cloud healthcare management service.

10.3.2.1 Cloud

When developing a cloud-based application, all of the code and data are saved and run in the cloud environment. Cloud-based applications were either built from the ground up in the cloud or shifted there from an on-premises infrastructure. It is possible to construct cloud-based applications by using these services because they provide abstraction from the administration, design, and scalability requirements of the underlying infrastructure of the cloud (AWS website).

10.3.2.2 Hybrid

It is possible to combine cloud-based resources with resources that are not stored in the cloud by using a hybrid deployment approach. The most common kind of mixed deployment is the one described above. Construction and enhancement of an on-premises architecture are possible, whereas cloud resources may be integrated with internal systems.

10.3.2.3 On-Premises

When virtualization and resource management systems are utilized to deploy resources on-premises, they are referred to as "private clouds" in many circles. On-premises installations remain popular despite the fact that they lack many of the advantages of cloud technology. This is due to the dedicated resources that they can provide. Application management and virtualization technologies are used in an effort to improve resource utilization in this deployment strategy, which is, in the majority of cases, the same as that of conventional IT infrastructure.

Therefore, making use of cloud computation can be partial or complete. This gives users choice and comfort to decide on which model to select.

10.3.3 Advantages of Cloud Computation

- With the usage of cloud storage in healthcare, electronic medical records may be maintained more efficiently.

  The American Recovery and Reinvestment Act of 2009 required the adoption of electronic medical records on January 1, 2014. When the American Recovery and Reinvestment Act of 2009 was passed by President Obama in February 2009, it contained provisions to assist the economy recover. To comply with the requirement, hospitals and healthcare institutions must show that they are utilizing electronic medical records to manage patient contact information. Medical cloud computing was designed to enhance the reliability, protection, and effectiveness of healthcare, as well as to better engage individuals and families, encourage care coordination, and protect the privacy and security of those getting medical treatment.

  Today, almost all of hospitals and healthcare organizations dropped paper record-keeping in favor of cloud computing for hospital management. Electronic health records (EHRs) stored in this technology may be versioned by healthcare professionals such as physicians, nurses, and others.

- The adoption of cloud-based healthcare storage makes it simpler for patients and caregivers to collaborate on patient care.

  As a result of the widespread usage of cloud-based electronic medical information storage, collaboration in patient care has become much more convenient. Cloud storage makes it possible for doctors to more readily see and exchange a patient's medical information in a collaborative environment, which is beneficial for both parties.

- Cloud computing assists in lowering the costs of data storage in the healthcare industry.

  Obtaining hard drives for data storage, as well as other IT infrastructure, will be required in order to guarantee that data are both secure and easily available while constructing on-site storage.

  It may be possible to decrease the initial costs of cloud-based healthcare systems so healthcare providers can devote their time and resources to what they do best: caring for their patients, increasing their overall efficiency.

- Greater data security in healthcare is made possible by cloud computing.

  A flood, a fire, or any other natural calamity might destroy all of your paper documents. The materials' lack of protection jeopardizes patient safety.

  As a result, healthcare practitioners may now use HIPAA-compliant cloud storage services to store and safeguard patient data. These services, which have access to patient electronic medical records, follow legally regulated data security and privacy requirements. Healthcare professionals have access to proper data storage options because of "the cloud."

TABLE 10.1
Comparison of traditional and cloud-based approaches

Dimension	Cloud Based	Traditional Approach
Capacity	Unlimited	Limited
Containers	Cheap rented storage	Local storage
Availability	24/7 over the Internet	Limited
Synergy	Real-time	Not real-time
Expenditure	Pay per use	Upfront cost and maintenance
Scalability	No limits	Limited
Accessibility	Anywhere, anytime	Limited

- Allowing for the development of Big Data applications on the cloud
  Because of the increasing usage of cloud-based data storage systems in the healthcare industry, patients' outcomes may now be improved with the use of "big data" apps.
- Cloud computing will have a positive impact on medical research.
  In the future, researchers in the medical field will greatly profit from the digitalization of healthcare data, which will be stored in cloud-based database storage, in a similar way to how cloud computing has benefited them.
  Cloud computing allows academics to get access to previously inaccessible computer resources, such as massive amounts of memory and processing power, that were previously unavailable. Table 10.1 shows the basic difference between cloud and traditional approach.

10.4 AMAZON WEB SERVICES

AWS, as a leading cloud service provider, has different modules that describe its working principles. Some of the important modules are discussed in this chapter for a clear view to cloud users.

10.4.1 COMPUTE

10.4.1.1 Amazon Elastic Compute Cloud

A "private cloud" is a term that is used to refer to on-premises deployments of resources that are made possible by virtualization and resource management systems. However, despite the fact that on-premises installations are less prevalent than cloud computing, they continue to be popular. This is owing to the fact that they have the ability to allocate resources to this cause. When deploying this approach, which is, for the most part, the same as when deploying traditional IT infrastructure, application management and virtualization technologies are utilized in an attempt to increase resource utilization and efficiency.

10.4.1.2 AWS Elastic Beanstalk

When you use AWS Elastic Beanstalk, you don't have to worry about the underlying architecture of your applications. AWS Elastic Beanstalk reduces administrative burdens without sacrificing control or flexibility. AWS Elastic Beanstalk takes care of all the complexities of introducing new features, balancing traffic, scaling, and monitoring the health of your application.

AWS Lambda, Lightsails, and Outposts are other popular compute resources available in AWS.

10.4.2 STORAGE

Amazon S3 (Simple Storage Service) is an object-centralized repository that contains data as objects within buckets. A file and its related information comprise an object. A bucket is a container for goods.

A bucket must be established first, with an username and an AWS region provided, before any data can be saved in Amazon S3. Then, on Amazon S3, simply upload your data as objects. Objects in a bucket are individually allocated a unique key (or key name) to distinguish them.

S3 features may be adjusted to fit your unique needs. For example, you may utilize S3 Versioning to save numerous versions of an item in the same bucket, which allows you to recover objects that are mistakenly deleted or overwritten.

Buckets and the things in them are confidential and may be accessed only if you specifically provide access privileges. You may use bucket guidelines, AWS Identity and Access Management (IAM) rules, access control lists (ACLs), and S3 Access Points to regulate access.

10.4.2.1 Amazon Elastic Block Store (EBS Volumes)

Using this service, Amazon EC2 instances may access block-level storage volumes. Like unformatted block devices, EBS volumes operate in the same manner. These volumes may be used as devices on your instances if you want to utilize them that way. EBS volumes that have been linked to an instance may be used as long-term storage, independent of the lifespan of the instance. Depending on their intended usage, these volumes may be used as block units or the basis for a file system (such as a hard disc or a hard drive). The variables in a volume may be adaptively updated when the volume is associated to an instance.

Amazon EBS are recommended for data that must be instantaneously accessible and require long-term durability. Any software needing fine resolution updates and improperly formatted block-level storage may benefit tremendously from the adoption of EBS volumes as the principal storage medium for file systems and databases. Amazon EBS is particularly suited to both data warehouses that depend upon unpredictable reads and writes, and to data rate apps that undertake long, consistent writes and reads.

10.4.2.2 Amazon Elastic File System (Amazon EFS)

Set-and-forget serverless Amazon Elastic File System (Amazon EFS) for AWS cloud services and on-premises resources delivers a basic elastic file system.

There's no need for monitoring or creating infrastructure to support expansion any more since it can scale up to petabytes on demand without disrupting applications as older storage systems can. Your organization can quickly and easily set up and manage data stores thanks to Amazon EFS's web services interface. When you utilize the service, you do not have to bother about establishing, upgrading, or maintaining a complicated file system architecture.

10.4.3 Amazon Machine Learning

Amazon machine learning service SageMaker is completely self-contained and does not require any additional software. With SageMaker, scientists and application developers can now create and train machine-learning-based models in a production-ready hosted environment, allowing them to collaborate more effectively.

There is no longer a need to maintain servers because it includes an integrated Jupyter writing notebook instance that allows you to gain immediate access to your information sources for the purposes of exploration and analysis. Also included are widely used machine learning methods that are well suited for large datasets and distributed computing environments. Because SageMaker's native support for bring-your-own-algorithms and frameworks allows you to tailor your distributed training to meet the specific workflows and requirements of your organisation, you can save time and money. Simple steps are required to launch a model from SageMaker Studio or the SageMaker console, and then deploy it in a secure and scalable environment with only just few keystrokes. Training and hosting are given prior to participation and require no upfront commitments.

10.4.4 Big Data Analysis in AWS

Amazon EMR (known as Amazon Elastic MapReduce) is a cloud computing platform that allows frameworks that are based on big data like Apache Hadoop and Apache Spark to be used to process and analyse huge amounts of data. With the help of these frameworks and their accompanying open-source projects, it is possible to do analytics and business intelligence activities. Massive amounts of data may be sent between Amazon EMR and other AWS data storage and systems, such as Amazon S3 and AWS DynamoDB, with no performance degradation. For handling big data, AWS uses EMR, which is shown in Figure 10.2.

10.5 CLOUD PRICING STRATEGY

Amazon provides one-year free tier trials. It offers many services that are more than enough to start a new venture. Different cloud providers' pricing statergies are studied and then compared in our study.

AWS began charging customers for EC2 Linux virtual machines and EBS volumes on a per-second basis in 2017. Per-second billing is still in effect for services, and it has been modified to some extent for other types of services as well. There are a lot of users talking about how the AWS services are billed per second for Windows on developer forums, despite the fact that there is no specific mention of

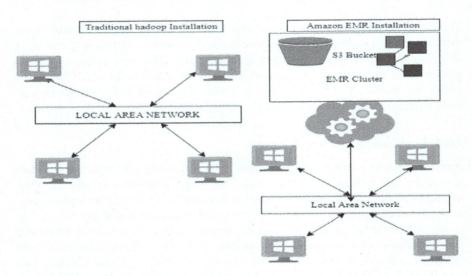

FIGURE 10.2 AWS storage and working.

such fees for RHEL or SLES on those sites. It's conceivable that there is an issue or that AWS is doing a test run, but neither of these possibilities has been verified by an official announcement from AWS as of yet. AWS has announced that for Linux instances, a user would be paid per second for each half instance that is used in an hour by a client. When it comes to per-second pricing, there is a 60-second minimum charge each transaction (Solanki and Kotecha, 2021).

Cloud computing systems designed for general purpose use as well as memory optimization are priced pretty much the same among Google Cloud and AWS.

Table 10.2 shows details of different cloud platforms available and the configuration of the instances available.

Prices for computer-optimized cloud instances on AWS and Azure are minimal when comparing the two cloud providers. The extensible computers and all-core turbo speed of Google Cloud Platform, on the other hand, make it the most costly service in this category (Solanki and Kotecha, 2021).

AWS and Azure's price plans for memory-optimized and highly fast instances are much more than Google Cloud's, owing to the fact that they do not offer 4vCPUs.

TABLE 10.2
Different cloud platforms and instances

Cloud Platform	AWS		AZURE		GOOGLE	
Instance Type	VM	RAM	VM	RAM	VM	RAM
General Purpose	t4g.xlarge	16 GiB	B4ms	16 GiB	e2-standard-4	16 GiB
Computer Optimized	c6g.xlarge	8 GiB	F4s v2	8 GiB	c2-standard-4	8 GiB
Memory Optimized	r5.xlarge	32 GiB	E4a v4	32 GiB	m1-ultramem-40	961 GB

TABLE 10.3
Pricing in different cloud platforms

Cloud platform	AWS		AZURE		GOOGLE	
Instance type	VM	Cost/hr	VM	Cost/hr	VM	Cost/hr
General Purpose	t4g.xlarge	$0.1997	B4ms	$0.198	e2-standard-4	$0.16
Computer Optimized	c6g.xlarge	$0.204	F4s v2	$0.212	c2-standard-4	$0.25
Memory Optimized	r5.xlarge	$0.302	E4a v4	$0.28	m1-ultramem-40	$6.3039

Alternately, they provide 40 and 12 virtual CPUs, respectively, as an alternative. Table 10.3 shows the pricing of the different instances in various cloud platforms. These prices are computed on a per-hour basis.

10.6 AWS – HEALTHCARE SOLUTIONS

Patient information must be kept secure and private in the healthcare industry, which is heavily regulated (Healthcare solutions, AWS). The healthcare industry must speed up innovation and maximize the potential of data, while still maintaining patient privacy and security. AWS offers the broadest and deepest cloud service portfolio, as well as function partner solutions, to the healthcare industry, allowing organizations to enhance patient outcomes and expedite data conversion.

Employees and industry specialists at AWS dedicated to healthcare have been working with healthcare organizations for the last 15 years to design and execute solutions with a single goal in mind: improving patient outcomes.

10.7 CONCLUSION

Cloud computing provides a broad variety of services and advantages, but there are still a number of difficulties that must be solved in order for the demand for this technology to continue to develop (Deyan and Hong, 2012). The most pressing challenges in cloud computing are data security, access control, and privacy protection. The risks connected with cloud computing must be kept to a bare minimum. Because of its extensive work in the area of data security, AWS is able to provide incredible cloud computing performance.

REFERENCES

Bali, S., & Singh, A.1. (2007). Mobile phone consultation for community health care in rural north India. *Journal of Ielemedicine and Ielecare*. 13(8): 421–424.

Blake, H. (2008, April). Innovation in practice: Mobile phone technology in patient care. *Br J Community Nurs*, 13(4): 160–165.

Buyya, R., Yeo, C. S., & Venugopal, S. (2008). Market-oriented cloud computing: Vision, hype, and reality for delivering IT services as computing utilities. In Proc. IEEE/ACM Grid Conf., 2008, pp. 50–57.

Chauhan, R., & Kumar, A. (2013, November). Cloud computing for improved healthcare: Techniques, potential and challenges. In *2013 E-Health and Bioengineering Conference (EHB)* (pp. 1–4). IEEE.

Deyan, C., & Hong, Z. (2012). Data security and privacy protection issues in cloud computing. In 2012 Internationals Conference on Computer Science and Electronics Engineering.

Dhilawala, A. (2019, March 5). *9 Key Benefits of Cloud Computing In Healthcare – Galen Data*. Galen Data. https://www.galendata.com/9-benefits-cloud-computing-healthcare/

Grogan, (2006). EHRs and information availability: Are you at risk? *Health Management Technology*, 27(5): 16.

Maria, A. F., Fenu, G., & Surcis, S. (2009). An approach to cloud computing network. In Proceedings of the 3rd International Conference on Theory and Practice of Electronic Governance, Bogota, Colombia, 10–13 November 2009; pp. 409–410.

Mohammed, A., & Maddikunta, L. (2014, July). Secured health monitoring system in mobile cloud computing. *International Journal of Computer Trends and Technology (IJCTT)* 13(3): 32–36.

Muir, E. (2011). Challenges of cloud computing in healthcare integration. Retrieved from http://www.zdnet.com/news/challenges-of-cloudcomputing-in-healthcare-integration/6266971 accessed July 23, 2013.

Perera, I. (2009). Implementing Healthcare Information in Rural Communities in Sri Lanka: A Novel Approach with Mobile Communication, *Journal of Health Informatics in Developing Countries*, 3(2): 1–15.

Solanki, J., & Kotecha, R. (2021, February 1). *Cloud Pricing Comparison 2021: AWS Vs Azure Vs Google Cloud*. Insights on Latest Technologies – Simform Blog. https://www.simform.com/blog/compute-pricing-comparison-aws-azure-googlecloud/ http://aws.amazon.com/what-is-aws/

11 Privacy and Security Solution in Wireless Sensor Network for IoT in Healthcare System

Rajesh Tiwari
Department of Computer Science & Engineering, CMR Engineering College, Hyderabad, India

Deevesh Chaudhary
Department of Information Technology, Manipal University Jaipur, India

Tarun Dhar Diwan
Chhattisgarh Swami Vivekanand Technical University, Bhilai, India

Prakash Chandra Sharma
Department of Information Technology, Manipal University Jaipur, India

CONTENTS

11.1	Introduction	150
11.2	Classification of WSNs Protocols	152
	11.2.1 Data-Centric Routing Protocol	152
	11.2.2 Multiple Path Routing Protocol	153
	11.2.2.1 Disjoint Path Routing Protocol	153
	11.2.2.2 Braided Path Routing Protocol	153
	11.2.2.3 N to 1 Multipath Discovery Routing Protocol	153
	11.2.3 Hierarchical Routing Protocol	153
	11.2.3.1 Initial Phase	154
	11.2.3.2 Neighbor Discovery Phase	154
	11.2.3.3 Clustering Phase	154
	11.2.3.4 Data Message Exchange Phase	154
	11.2.4 Routing Protocol Based on Location	154
	11.2.5 Mobility-Based Routing Protocol	154
	11.2.6 Quality of Service-Based Routing Protocol	155

DOI: 10.1201/9781003217091-11

11.2.6.1 Sequential Assignment Outing (SAR) Protocol 155
11.2.6.2 SPEED Protocol 155
11.2.6.3 Quality of Service – Aware and Heterogeneously Clustered Routing Protocol (QHCR) 155
11.3 Privacy and Security Issues in WSN 155
    11.3.1 Security and Privacy Issues 156
        11.3.1.1 Denial of Service Attack 156
        11.3.1.2 Manipulating Routing Information 157
        11.3.1.3 Sybil Attack 157
        11.3.1.4 Sinkhole Attack 158
    11.3.2 Clone Attack 159
    11.3.3 Selective Forwarding Attack 159
        11.3.3.1 HELLO Flood Attack 159
11.4 Security and Privacy Solutions 160
    11.4.1 Use of Effective Key Management 160
    11.4.2 Use of Efficient Public Key Infrastructure 161
    11.4.3 Effective Use of Multiclass Nodes 161
    11.4.4 Efficient Clustering of Modules to Increase Safety of WSN ... 162
    11.4.5 Point-to-Point Protection Approach 162
    11.4.6 Registration and Key Management Phase 163
    11.4.7 Secure Data Exchange Phase 163
    11.4.8 Generating Perturb Phase 163
    11.4.9 Signature and Perturbation Phase 164
    11.4.10 Authentication Phase 164
    11.4.11 Decryption and Authentication 165
11.5 Conclusion 165
References 166

11.1 INTRODUCTION

As seen in Figure 11.1, a wireless sensor network (WSN) is a portable synchronized network consisting of autonomously configured components which use sensing devices to accurately measure the situations in its location [1]. WSNs have been included across a number of conditions, such as healthcare, public safety and agricultural management, mapping, armed forces support, and interference prevention, among others. Protection in WSNs is becoming more important, not because of a lack of effective security features, but although because of the distinctive features of WSNs, many current systems were inefficient. That is, the computational ability and equipment available to WSN entities are minimal. Sensors in WSNs can share information between each other, however its basic function is to hear, collect, and calculate information. This information is transmitted to sink, that it can be used or transmitted among several systems, via multiple hops. WSNs require reliable network architecture to accomplish convenient interactions [2]. They help WSNs communicate by determining the best paths for information transfer and maintaining those routes for future transfers. Because of the diversity of WSN sensor node, numerous standards for various WSNs have been created, based upon the type

Security based Sensor Network in Healthcare

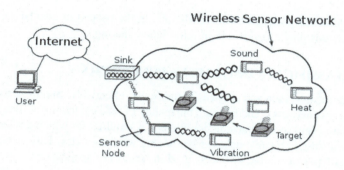

FIGURE 11.1 WSN in practice.

of the sensor node as well as the program. For example, MWSNs have their own methods, whereas SWSNs have their own methods.

In WSN, there are two methods of communication: single hop and multi hop. In single hop, the entity transmits the information toward destination in single step. In the meantime, sensor nodes in WSNs might depend on each other to transmit data to distant terminals. Multi-hop transmitting is the name for this form of communication. Multi-hop communication is a technique in which data is transferred among starting point and target points with the help of intermediate points. It improves the efficiency of WSNs by enabling energy-drained nodes to pass information to the target node via neighboring nodes across the routing track [3]. Multi-hop communication is concerned with a number of security and privacy concerns. Spying, sinkhole, misuse Sybil, copy, blackhole, spoofing, and other problems impact the WSNs' credibility, availability, and data confidentiality.

Researchers are continually searching for new ways for providing efficient healthcare facilities to improve patient outcome. Wireless sensor nodes connected over network are transforming the way information is being transmitted in between nodes. These sensors help collecting the patient-related records and storing over a centralized database system, thus providing instant access for medical staff and improving patient–staff communication. The live patient data are available on run-through digital health systems that improve patient treatment. As patient data are highly sensitive and personal, the systems must ensure the accuracy of data collected, and patient confidentiality should not be altered at any point.

Numerous security mechanisms for WSNs have been suggested; although, due to sensor limited resources, some of these protection approaches are unsuitable for WSNs. As a result, their acceptance in WSNs is unlikely. This is due to the topology among most WSNs being unstable. Unlike other networks, some WSNs have wireless terminals that alter the network topology on a regular basis, rendering it difficult for such a wireless network to be using current protocols designed for static nodes. In addition, vast amounts of data are transmitted over WSNs, which enhances demand on the WSN's cellular networking networks. All of this demonstrates that WSN protection and secrecy technologies should not only be lighter in aspects of technical, connectivity, as well as resource expenses, and moreover promote convergence and multi-hop to minimize load and prolong network's lifetime. In the context of healthcare, WSN and IoT applications play a major role in information sharing.

11.2 CLASSIFICATION OF WSNs PROTOCOLS

To transfer medical data and services securely and effectively, minimize the stress on wireless access networks, and increase the quality of medical treatment such as surgery, a collaborative and secure transmission method is required for which a reliable routing protocol will sustain reliable data transfer among wireless nodes. This section provides the various types of routing protocols used in healthcare applications.

11.2.1 DATA-CENTRIC ROUTING PROTOCOL

At a given path, the data-centric routing protocol integrates records from different sensors. Until transmitting records to root node, this avoids duplication and reduces the overall volume of data transfer. Data-centric routing protocols include guided diffusion, gossip routing process, and the sensor-related protocol for transferring the information through SPIN [4] protocol.

It is a data-centric protocol for WSNs that is built on negotiations. Every entity uses metadata to identify its data, and a sensor uses its meta-data to negotiate. As a result, each node will decide whether to transmit records, eliminating duplication data transfer across the system. The sender transfers its information after the agreement, as seen in Figure 11.2; point A begins by transferring its hop message to its immediate nearby point B. When this query is authorized, node A transfers its information to node B, which further repeats the process. This process is done to locate a nearby node and hop the data to that point before it achieves its destination. Since each node only executes a single hop, the SPIN protocol saves time and money [5]. WSNs are protected from flooding attacks by SPIN's hop demand and acceptance data packet. While the SPIN protocol is best suited to networks for no downtime, it might be utilized in satellite or downlink connections.

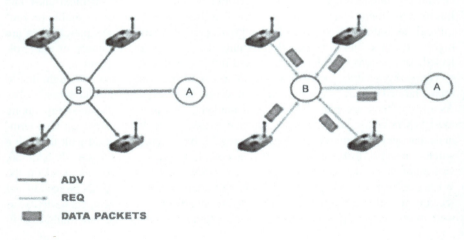

FIGURE 11.2 SPIN protocol.

11.2.2 MULTIPLE PATH ROUTING PROTOCOL

It creates a multiple path (Initial and backup route) from supplier to recipient point for successful data transmission. In the event that the primary route fails, the backup route is used. Fault tolerance is reached as a result of this. Even though, the expense of managing numerous routes among both the sender and receiver raises the packet forwarding expense. Multipath-based routing come in a variety of flavors. A few of them are as follows:

11.2.2.1 Disjoint Path Routing Protocol

Any source node in a disjoint route routing protocol searches for the shorter disconnected multipath to sink device. It distributes the load equally between all disconnected routes. All the routes throughout this multipath [5] distribute no sensing module. The procedure is efficient, but it has a lot of maintenance and uses a lot of resources.

11.2.2.2 Braided Path Routing Protocol

The protocol initially chooses the initial route; after that, the better route from sender to sink module is selected for each sensor. However, this route would not involve the default point. Idealistic braided multipaths are the perfect alternate routes that aren't inherently separate from its primary routes. These alternate routes are either on or very similar to the main path, implying that the power demand on both the preferred and substitute routes is nearly equivalent [6].

11.2.2.3 N to 1 Multipath Discovery Routing Protocol

This routing protocol uses a flooding-based algorithm for transferring the packets. This protocol is an example of segment-by-segment routing (SSR). Multiple node-disjoint routes are identified as well as autonomously managed using the MSSR protocol, which separates a single path into various path. This protocol successfully handles overcrowding as well as eliminates it.

11.2.3 HIERARCHICAL ROUTING PROTOCOL

Network points are divided into various levelled clusters by the hierarchical routing protocol. The algorithm chooses a terminal with the higher residual energy as cluster head for each cluster. The sensor readings of every cluster unit is passed via the cluster heads of a channel [7]. Before transmitting the details to the sink, the centroid entity aggregates the sensor information from all the entities on the network. The multi-hop transmitting method of the hierarchical routing lowers power intake. Additionally, data consolidation conducted by the cluster head eliminates network load. Hierarchical routing protocols include LEACH, TEEN and APTEEN, and SHEER. TEEN provides excellent results by reducing the amount of transmitting data. SHEER was introduced by the author in [8]. In WSN, it employs an evolutionary probabilistic transmitting system to determine the best path. For key delivery, verification, and secrecy, SHEER uses the hierarchical key establishment scheme (HIKES). SHEER is classified into four stages, as follows:

11.2.3.1 Initial Phase

- The base station (BS) calculates the key $K_R = HMAC(I_R \| O_R)$, produces a broadcast access code N_R as well as encodes it as $N_R^l = Enc_{K_R}(N_R)$. The BS pre-loads all sensor module with N_R^l and retains I_R and O_R.
- The set-up call is transmitted on BS as $Nb \| IR \| OR \| Enc_{K_R}(init \| N_b \| N_R \| N_R^{ll})$, in which init is perhaps startup call, O_R is the index, O_R is the offset of K_R, and N_b is time stamping produced by BS.
- On gathering the initialization letter, the sensor module captures then decodes $Enc_{K_R}(init \| N_b \| N_R \| N_R^{ll})$, regenerates N_R^l, and matches it from N_R in the starting letter got if both look alike, the BS has been effectively verified. Then, it swaps N_R into N_R^l with in new arrival, activates the timers, then moves onto another step.

11.2.3.2 Neighbor Discovery Phase

The sensing entities create its hopping connection from their neighboring entity on the time of the neighbor discovering process. From hearing to transport phase, each entity turns. In hearing phase, the node transmits a Hey msg, including its identification, a sequence number, and an encoded heading usually containing the sensor keys before it receives a response by its neighbors.

11.2.3.3 Clustering Phase

Centered on several criteria, a cluster with a specified amount of nodes from a cluster head is chosen throughout this step.

11.2.3.4 Data Message Exchange Phase

The cluster heads transmit information from every sensor towards the base station. This centralizes data transfer, which prevents cluster overlap.

11.2.4 Routing Protocol Based on Location

The data are routed using the location-depending routing protocol based on the distance between the supplier and recipient points. It determines approximate routing energy by calculating the distance between source and destination nodes. Shruti [9] suggested a routing protocol based on venue. The protocol calculates their distance based on the frequency of the received signals. To save resources, all non-active nodes are placed into sleep mode according to their protocol. In location-based requests, the location of sensing modules is used to guide the request from BS to the case. The network will choose the right route based on the user's location. GAF protocol for MANET is another instance of a location-based protocol. GAF saves resources and lowers transmitting overhead, making it ideal for WSNs, LAR, EELAR, GLAR, and many other location-based protocols.

11.2.5 Mobility-Based Routing Protocol

The MBRP is a simple protocol that guarantees the information to be transferred from sender to recipient points. MBRP include TEDD, SEAD, TTDD, and data MULES.

Such routing protocols operate along the channel's dynamism in terms of topology. The unit nearest to the sink entity transmits more than anyone, reducing its lifespan sooner than others. The protocol suggested by [10] the other case of a MBRP. For WSNs, the authors suggested an algorithm-related temperature-aware mobility. The algorithm uses a store-and-pass mechanism to address the complexities of human posture flexibility. In this protocol, routing data packet were placed in a short-term storage termed as buffer. The buffer reroutes missing records to every intermediate module that has lost communication among the sender module for an extended period of time. Temperature is often used in their protocol to evaluate the intermediate node.

11.2.6 Quality of Service-Based Routing Protocol

This set of rules blends efficient data transmission to the sink module with several prespecified metrics related to QoS. The following are several current QoS-based routing protocols:

11.2.6.1 Sequential Assignment Outing (SAR) Protocol

To achieve successful data transmission, the SAR procedure uses resources, QoS on every route, as well as the packets preference standard as QoS metrics. For successful data transfer, the SAR protocol finds via its sink module to their sensor terminals, several routes are used. During data transmission, the SAR protocol acknowledges power conservation and faults responsiveness as well as minimizing the average measured QoS metric [11].

11.2.6.2 SPEED Protocol

SPEED is another instance of a quality-of-service-based routing protocol. In SPEED, each sensors saves details about its neighbors in terms of improving the protocol's efficiency. Congestion avoidance mechanisms, for example, are used in the SPEED protocol to prevent overcrowding. The function is based on the information provided by the nodes. The stateless regional non-deterministic forwarding (SGNF) routing framework in SPEED collaborates with four network layer components. The amount of energies required for transmitting data in this protocol is beyond comparison to the networking application's efficiency.

11.2.6.3 Quality of Service – Aware and Heterogeneously Clustered Routing Protocol (QHCR)

It's a low-power networking solution for latency-sensitive, bandwidth-intensive, time-critical, and QoS-aware programmes that are employed by heterogeneous WSNs. The QHCR protocol enables specialized routing for real-time and delay-sensitive programmes at a low energy cost. The QHCR procedure involves three phases: knowledge gathering, cluster head placement, and intra-cluster coordination.

11.3 PRIVACY AND SECURITY ISSUES IN WSN

As most of the data transmission in healthcare system is carried over wireless network, it requires a secure data transmission over channels. The chances of losing data that are collected over sensor nodes embedded in various healthcare devices is very high and can occur due to various reasons along the transmission paths. Due to the resource constraints, the majority of current WSN routing protocols and authentication suggestions are incompatible with WSNs [12]. Depending upon the restrictions such as resource constraints, channel bandwidth, sensors, etc. different types of security measures are considered. This segment covers a variety of security problems and their remedies.

11.3.1 Security and Privacy Issues

With the time, the need of information transfer over WSN is grown. Healthcare records are very sensitive and particularly are attractive targets for cybercriminals, since they hold the vital information required for personal identity theft, social engineering, financial medical fraud, tax fraud, and medical insurance fraud. To solve their limitations, most WSNs use a multi-hop transmission mode. During hopping, attack on the sender data and nodes' identity are a big issue with multi-hop transmission. In a resource constrained WSN with a sender entity transmitting information to the target by multiple intermediate entities, there may be a chance of intrusion, identification stalking from a hacker, relaying, and misuse of source data by any middle module. WSNs are known to operate in hostile environments and are vulnerable to side channel attacks such as differential power analysis. In order to extract the concealed key or disrupt the system, the attacker monitors the machine, executes the identical technique, and meticulously notes the amount of energy spent on a cycle-by-cycle basis. Scalar blindness is commonly used for protection technologies focused on cryptography to avoid this. Int m is used to blind the scalar multiplication, whereas m is the sequence of the point $P \in Eq$, such that $mP = 0$.

For instance, in place of evaluating $Q = kP \bmod q$, $Q = (k + m)P \bmod q$ is computed.

Another problem in WSNs is how to protect the identity of sender and receiver points during multi-hop from the prying eyes of intermediate nodes and adversaries. That is, the packet between the sender and receiver points must have some kind of minimalist identity verification feature(s). Other WSN attacks are listed further down.

11.3.1.1 Denial of Service Attack

This kind of attack targets the sensors network weaknesses by attempting to tamper with the network system. Legitimate users are unable to access resources due to a denial of service (DOS) assault. In safety-critical networks, this sort of assault might be detrimental to the network's operation. One of the strategies used by adversaries to carry out DOS attacks is to try to flood the system with messages in order to increase undesired traffic. Filtering received data according to its materials and identifying companies with a significant number of faulty email may help identify a DOS assault. Testing for inconsistencies in msg transmitted by adjacent entities detects unreliable messaging [13]. Figure 11.3 illustrates a DOS attack.

Security based Sensor Network in Healthcare

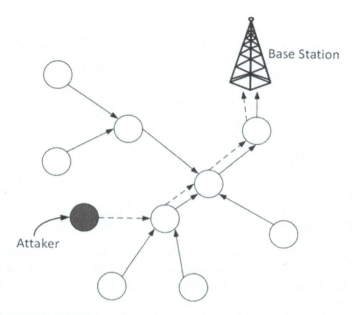

FIGURE 11.3 Denial of service attack.

11.3.1.2 Manipulating Routing Information

The networking details among multiple sensing devices is really the object of this strike. It is possible to activate it by sending fake or retransmitted data packets. Attackers have the ability to construct forwarding patterns, remove or reject network activity, and stretch or reduce origin paths may be able to complete it. It is a passive strike that is trouble-free to begin, however tough to track. However, a distinct identity for the chosen direction may be formed. (Any attempts to register packets through one point and re-tunnel them at another location are monitored from the base station while comparison the enriched path identification with hashing of all the appended pseudonyms or identifiers of every entities embroiled in the multi-hop would be discovered by the access point while comparison of patterned route identification with checksum of all the attached identifiers or identity of all entities participating throughout the multi-hop).

11.3.1.3 Sybil Attack

In these attacks, the intruder hacks the WSN by impersonating other people to interrupt network protocols. A Sybil attack will result in a service denial. Because a Sybil entity generates unauthorized identities to tear down the individual modelling of every entity, it can have an impact on routing. Sybil is widely used in peer-to-peer networking and is also used in WSN [14]. Furthermore, identifying and fighting against Sybil attacks is substantially more challenging due to WSNs' poor energy and computational capacity. Various techniques were designed to counter Sybil's onslaught in WSN. For example, a pair-wise key-based recognition system imposes a restriction on the number of identities that one entity may use [15]. However, it demands pre-assignment of secret key to sensors. Figure 11.4 illustrates sybil attack.

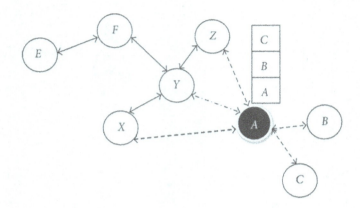

FIGURE 11.4 Sybil attack.

The other method to overcome the Sybil attacks is to verify the identity of each transmission point. This may be achieved in a responsive or constructive manner. Even before to data transfer, a sensor should have sufficient identifying criteria to distinguish itself between several existing sensing modules. A resources testing is the most popular process. An alternative method to enhance the value of identities generation in comparing to the worth. A Sybil attacker's goal is to get as many identifications as possible. Raising the cost of creating an identity and lowering the prospect of having numerous identifications will foil Sybil strikes. To avoid a Syb assault, a detectable reference number and network-node identities created by the access point are often employed.

11.3.1.4 Sinkhole Attack

This infiltration prevents the sink node (base station) from obtaining complete and accurate data from multiple sensors, jeopardising higher layer programmes. An adversary utilises this attack to maintain itself attractive to its neighbours in the hopes of directing more traffic to it [16]. As a consequence, the intruder attracts all of the congestion intended for the sink module. After that, the attacker will launch a more severe network attack, such as packet routing, alteration, etc. Because its entities communicate information to its root terminal for the majority of the time, WSN is particularly vulnerable to infiltration. Figure 11.5 illustrates a sinkhole attack.

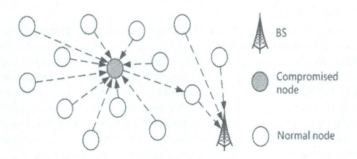

FIGURE 11.5 Sinkhole attacks.

Security based Sensor Network in Healthcare 159

In a cloning attack, the intruder initially targets and catches appropriate sensing entities in WSNs, then extracts all of their data from their memory, replicas it onto several sensing terminals to establish replica nodes and eventually distribute it to the channel. After cloning a node, the attacker may initiate some other strikes. This assault can be detected in two manners: clustered and dispersed techniques. The centralized technique employs a sink module to identify and sanction replicate module behaviors, whereas the decentralized method involves chosen entities to identify and sanction replicate module events in the channel [17].

11.3.2 Clone Attack

In a cloning attack, the intruder initially targets and catches appropriate sensing entities in WSNs, then extracts all of their data from their memory, replicas it onto several sensing terminals to establish clone nodes and eventually deploys it to the network. After cloning a node, the attacker may initiate some other strikes. This assault can be detected in two manners: clustered and dispersed techniques. The centralized technique employs a sink module to identify and sanction replicate module behaviors, whereas the decentralized method involves chosen entities to identify and sanction replicate module events in the channel. Since distributed approaches use device positioning data to distinguish copies and sensing entities using common identities but distinct locations, distributed approaches are ideal for stable WSNs. In mobile WSNs [18], though, it's a whole different storey; sensors modules are constantly shifting their positions, and these nodes are constantly entering and exiting the system. As a result, node positioning data are not thought to be the perfect method for identifying duplicate modules. The mentioned attacks may be launched by a clone node.

11.3.3 Selective Forwarding Attack

WSN routing techniques based on multi-hops assumed that the arriving packets should be re-hopped by practically all nearby entities. Malicious modules preferentially forward such packets, while others are deleted. Selective forwarding attacks are often effective when the invader is actively involved in the data exchange.

11.3.3.1 HELLO Flood Attack

The relation among entities is violated in this attack. To introduce itself to their neighbors, many routing protocols demand sensing modules to transmit HELLO packets. An intruder could use this to trick sensor nodes into assuming they were in radio transmission range of the sender node when they aren't. The researchers suggested a new distance-based approach for identifying the HELLO flood attack in [19]. Figure 11.6 illustrates HELLO flood attack.

Nodes equate the RSS of the sent HELLO packet with threshold width, as well as the node's range to the chosen cluster head (CH). Then only nodes whose RSS and range fall below the defined thresholds are permitted to enter the network. CH, for instance, transmits their own position information during the initialization process of the LEACH protocol [20].

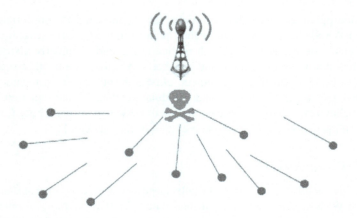

FIGURE 11.6 HELLO flood attack.

The modules getting HELLO packets from CH calculate the distance (Dist), which is described down:

$$Dist\,[sqrt\,[sq\,(x2 - x1) + sq\,(y^2 - y^1)]]$$

Here, $(x1;\ y1)$ are the sensing node's positions for acquiring data packets, and $(x2;\ y2)$ are CH locations. Every sensing entity calculates the radio signal strength value (RSS) and range among (Dist). It is utilized to figure out which cluster the entity belongs into, i.e. if *(RSS < ThRSS and Dist < ThDist)* then the entity is labelled as a "mate of cluster," or it is not a mate of cluster.

11.4 SECURITY AND PRIVACY SOLUTIONS

WSN framework has currently gotten a lot of publicity, which has resulted in new safety problems and architecture concerns [21]. We addressed related experimental efforts throughout this portion on the implementation of protection mechanisms for WSNs using numerous techniques like efficient key management, public key infrastructure, multiclass nodes, and node classifying to enhance the protection of routing protocols in WSNs.

11.4.1 Use of Effective Key Management

Du et al. proposed a framework that included an explanation of good key management. The moderate sensors in the diverse channels are used in their framework. The key distribution strategy's security evaluation and performance analysis suggest that it offers greater protection with little complexity than traditional key managing systems. According to the protocol, the L-sensor has a few pre-assigned keys, and every H-sensor has a few pre-assigned keys. The H-sensor is selected because it is more resistant to manipulation and has more storage than the L-sensor.

They adopt an asymmetric pre-distribution key distribution technique since the number of pre-allocated keys in an H-sensor and an L-sensor vary [22].

11.4.2 Use of Efficient Public Key Infrastructure

Yu in solved the WSN safety issue by applying public key encryption as a method to guarantee the sink node or base station's authenticity. The approach is separated in two stages: the entity to sink handshake, wherein sink and sensing entities generate a key pair for trustworthy data sharing, as well as the sink-to-sink handshake, wherein sink and sensing entities create a secret key for trustworthy data sharing. During stage 2, the secret key should be utilized for encryption of data. Their method is simple to apply and needs little computing power. Their scheme's only drawback is that before data can be exchanged, all of the network's participants must agree on a shared key. On the other perspective, every strategy relying on a unique key is susceptible towards key theft, i.e. when a device is tampered with, not just the common key under danger, but also the entire network. Chan et al. [23] have proposed a PKI-based solution for safe key exchange in WSNs. Their method offers a primary control system for WSN implementations that can accommodate sink mobility and reliably deliver data to nearby sensors and sink. They also demonstrated data authentication encryption and a tool for detecting and thwarting DoS attacks.

11.4.3 Effective Use of Multiclass Nodes

The two-tier protected routing (TTSR) protocol proposed by Du et al. [24] is a modern secure routing protocol for heterogeneous sensor networks (HSNs). Both intra-cluster and inter-cluster routing systems are used in the TTSR protocol. For data forwarding, intra-cluster routing creates a minimum spanning tree (shortest path tree) among L-sensors in a cluster. In the case of inter-cluster routing, data packets are sent from the source node to the sink node through H-sensors in the relay cells. Because of its tree-based routing and relay over relay cells, TTSR is subject to spoofing, selective forwarding, sinkhole, and wormhole attacks.

[25] introduced an innovative QoS routing protocol for MANETS that involves range measurement and slot reservation. Their QoS routing protocol took the benefit of multi nodes' multiple transmitting capabilities. Three secret keys were included in their protocol:

1. A shared key known by the sink as well as all other entity.
2. Node secret key distributed by two different neighbors as well as renewed in the path exploration process.
3. Among the module as well as the sink unit, there is a shared primary key.

Transmitted data are divided into multiple data sliced by the QoS routing mechanism.

All slices are linked via a specific path of the identified multi-path.

11.4.4 Efficient Clustering of Modules to Increase Safety of WSN

The dominant nodes evaluate the sensor data privately and generate the authorized recommendation for the destination node in a group-based WSN protection scheme [26]. In this category, sensor nodes are grouped into smaller clusters, with each cell having a specialized sensor node responsible for relaying multi-hop packets. Figure 11.7 illustrates the system model of WSN.

As a result, network division of labour is feasible, resulting in a low-power system. In [27], researchers proposed a group-based protection mechanism for distributed WSN, which consists of three components: N numbers of common sensing elements, Y numbers of cluster dominator nodes, and one or maybe more sink nodes.

11.4.5 Point-to-Point Protection Approach

A point-to-point security mechanism includes stabilized networking for any pair of terminals across this multi-hop path. We thoroughly explain a standard P2P protection approach for multi-hop supported WSNs introduced in [28,29] to demonstrate the architecture and effectiveness of point-to-point solutions. To address protection and privacy concerns in WSNs, they suggested an efficient point-to-point encryption scheme that uses a point-to-point shared mechanism of identification, perturbation, and pseudonym. They used elliptic curve encryption, hashing algorithms, and XOR operation to establish an optimal protection strategy for distributed WSNs that reduced computation costs and power usage. PoP security system consists of the registering and key distribution processes, secured information exchange, perturbation formation, biometric and masking, identity management, and authorization and decoding processes.

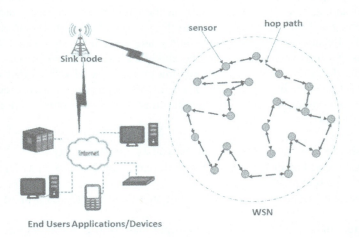

FIGURE 11.7 System model of WSN.

11.4.6 Registration and Key Management Phase

Every modules's identifier ψ is transmitted to BS. The following is how BS produces a specific pseudonym and device identification:

Step 1. BS randomly produces s, $\rho \in Z_q^*$, since its master private key pairs, and calculates and shares its public parameters $\varphi = (\rho + \mu)P \, mod \, q$, here P is the elliptical curve's producer Eq and q is the order of E.
Step 2. Every module i picks a distinct $r_i \in Z_q^*$ at irregular intervals, calculates its two-way transmission factor β_i as $\beta_i = (r_i + \mu)P \, mod \, q$, and sends it β_i to the rest of the system.
Step 3. BS then calculates N_i as $N_i = H(\rho \oplus \psi i)$ and pseudonym F_i as $F_i = H(N_i \| s \| \psi i)$ for all enrolled nodes. It takes the module i's distributing factor βi in accordance to calculate its node-base session public key $\gamma_{bs \to i}$ as $\gamma_{bs \to i} = \rho \beta_i$ and delivers the symmetrically encoded node's F_i & N_i to module i as $E_{\gamma_{bs \to i}}(F_i)$.
Step 4. Every entity produces its associated node-base station public key as $\gamma_{i \to bs} = r_i \varphi$ after receiving its encoded pseudonym and applies this to decode the acquired encoded pseudonym.

11.4.7 Secure Data Exchange Phase

The primary SN signatures M as well as creates perturb to safeguard M in order to transfer info M. Then, it uses its node-destination public key $\phi_{sn \to ds}$ for encrypting that obfuscated msg packet as σ. The signature δ, perturbed data Pp, pseudonyms of the primary sender module F_{sn}, and target module F_{ds} are all included in the msg packet σ.

11.4.8 Generating Perturb Phase

It imposes a first degree of data protection. It's utilized to get rid of semantic patterns created by a lot of variances in the results. To perturb the data M, the perturbation employs a new [30,31] additive noise production process. Initial sender and receiver modules autonomously produce a sequence of perturb λ for session τ as shown below:

i. The SN and its target module generate the perturbations factors α_{sn}, α_{ds} by picking a distinct $m_1 \in Z_q^*$ & $m_2 \in Z_q^*$ at random and calculate $\alpha_{sn} = (m_1 + \mu)P \, mod \, q$ & $\alpha_{ds} = (m_2 + \mu)P \, mod \, q$, correspondingly.
ii. SN calculates perturbation seed ϑ as $\vartheta = m_1 \alpha_{ds}$ by applying the target perturbation factor α_{ds} for session.
iii. SN creates the perturbation sequence as $\lambda = \{\lambda_1, \lambda_2, \lambda_3 \ldots \lambda_k\}$, where

$\lambda_1 = H_\vartheta(\vartheta \| F_{sn})$, $\lambda_n = H_\vartheta(\lambda_{(n-1)})$ for $n = 2 \ldots k$. For sessions and target module of pseudonym F_{ds}, remove every perturbation factor of perturb value n 1 in its storage. It substitutes the old, encoded perturbation factors with newer ones, i.e. $[(\lambda_{n-1} \| m_1 \| n \| F_{ds}) \oplus \vartheta]$ replaced with $[(\lambda_n \| m_1 \| n \| F_{ds}) \oplus \vartheta]$.

iv. By reiterating step (iii) utilizing the earlier utilized perturb λ_{n-1}, Initial SN calculates a fresh perturb for each new packet transfer along the same event (n − 1). SN, on the other hand, produces a new ϑ for a fresh session and target module by using the procedures (i)–(iii).

11.4.9 Signature and Perturbation Phase

The packets are signed and perturbed by the primary sender module using the mentioned procedure [32]:

a. The source-destination session public key $\phi_{sn \to ds}$ is computed by both the SN as well as the target modules.
   i. K_1 and k_2 are generated in a distinctive way by SN and target modules, respectively.
   ii. SN calculates $\phi_{sn \to ds}$ as $\phi_{sn \to ds} = \kappa_1 \beta_{ds}$ by extracting the two-way distribution factor of the target module β_{ds}.
b. $P_p = M + \lambda_n$ Perturbs M by signing its info M with its sending-receiving session public key $\phi_{sn \to ds}$ as $\delta = H \phi_{sn \to ds}$.
c. To maintain 2nd-tier data secrecy and truthfulness of the messaging and transmission data, SN produces its msg packets $\sigma = \delta \| P_p \| F_i \| F_j \| n$ and encodes it as $\sigma_\sigma = \sigma \phi_{sn \to ds}$, while F_i and F_j are the pseudonyms of both the sender and target modules, respectfully.
d. Prior to hopping P_p to the IN, SN executes PoP verification with its IN, as detailed on the coming portion.

11.4.10 Authentication Phase

After the signature as well as perturbation process, the sender module begins the PoP verification with the IN as shown below:

i. SN produces a verification identifier ω and current time t_s.
ii. SN & IN randomly produce $\upsilon \in Z_q^*$ and $\varepsilon \in Z_q^*$, subsequently. SN evaluate its PoP variable for verification as $n_{sn} = (\upsilon + \mu)P \bmod q$, while IN computes its own as $n_{in} = (\varepsilon + \mu)P \bmod q$ and transmits that to SN, that calculates the PoP event verification code $\phi_{sn \to in}$ as $\phi_{sn \to in} = \upsilon \cdot n_{in}$.
iii. SN afterward encodes the concatenated verification identifier ω, pseudonym of sender, pseudonym of IN, and current time as $E_{\phi_{sn \to in}}(\omega \| F_{sn} \| F_{in} \| t_s)$, concatenates it among n_{sn} as $E_{\phi_{sn \to in}}(\omega \| F_{sn} \| F_{in} \| t_s) \| n_{sn}$, and transmits it to its IN.
iv. On receiving $E_{\phi_{sn \to ds}}(\omega \| F_{sn} \| F_{in} \| t_s) \| n_{sn}$, IN extracts n_{sn} then computes its $\phi_{in \to sn} = \varepsilon \cdot n_{sn}$. It decrypts the received $E_{\phi_{sn \to in}}(\omega \| F_{sn} \| F_{in} \| t_s)$ using its $\phi_{sn \to in}$ to capture ω and t_s. It then uses $\phi_{in \to sn}$ to re-encrypt the collected ω and t_s before sending it in return to SN. SN decodes it utilizing its $\phi_{sn \to in}$ and validates it by matching it to the actual values of ω and t_s. If the two values are equivalent, SN hops its encoded data packet σ_σ. The IN

afterward will be a short-term SN and reiterates this process with the IN it has chosen until the data reach the target point t.

11.4.11 Decryption and Authentication

The mentioned steps are used by the target module to retrieve and verify the obtained packet M:

i. The target module calculates the target of used perturb P by extracting the two-way distribution metric of SN and β_{sn}.
ii. By evaluating the valuation on the perturb dataset n, the target module reproduces utilized perturb λ_n^l. When $n = 1$, the sender is novel to target module, and the target module then goes through the perturbation production process to get the perturb seed that will be utilized to recalculate the old perturb. If n is greater than one, the event is for an older target module [30]. By conducting stage three of the perturbation production process, the target module retrieves the encoded previous perturb for the sender from its storage, decodes it, then utilizes it to gain utilized perturb. It retrieves the n msg using unperturbed P_P as: $M^l = P_P - \lambda_n^l$.
iii. Receiver module authenticates the sign by re-signing the open label msg M^l using the $\phi_{ds \to sn}$ as $\delta^l = H\phi_{ds \to sn}(M^l)$. If $\delta^l = \delta$, the target entity adopts the records if the perturbation, data, and sender entity are all accurate; or else, data would be rejected. Encode perturbation metric as $\lambda_n \oplus \vartheta$, $m_2 \oplus \vartheta$, $n \oplus \vartheta$, F_{sn}. Remove all earlier encoded perturbation metrics saved in storage, F_{sn} and swap it $(\lambda_n \| \vartheta \| m_2 \| \vartheta \| n)$.

11.5 CONCLUSION

This chapter gives an outline of WSNs, as well as the protection and privacy policy that surrounds them. The chapter covers in-depth knowledge of WSN confidentiality and protection concerns in healthcare applications. There is a discussion of several present works in WSN routing protocols used in various applications and services related to medical and healthcare. The chapter also aids scholars in gaining a better understanding of existing WSN routing protocols and safety mechanisms used in medical-related applications. We assessed methodologies as well as testing activities utilizing sensors and QoS elements. One of the most important issues is to meet applications' QoS needs while also offering advanced abstractions for WSN safety in transmission of data collected through sensor nodes. Such tests may be readily executed in parallel with suggested methodology by broadcasting wirelessly to a large number of nodes at the same time. As a result, the suggested method may be applied to a wide range of testing situations. A safe model is suggested that uses QoS to provide a flux of protection classifications that provide varying degree of protection and integrity in the medical field. However, we discovered that the impact of evaluation criteria creates a significant strain on the entire platform's QoS, resulting in decreased speed.

REFERENCES

[1] O. Olufemi Olakanmi and A. Dada. 2020. Wireless Sensor Networks (WSNs): Security and Privacy Issues and Solutions. 10.5772/intechopen.84989

[2] L. Eschenauer and V. D. Gligor. November 2002. "A Key-Management Scheme for Distributed Sensor Networks." In Proceedings of the 9th ACM Conference on Computer and Communications Security (CCS'02), 41–47, Washington DC, USA.

[3] R. Chandrakar, R. Raja, R. Miri, U. Sinha, A. K. S. Kushwaha, and H. Raja. 2022. "Enhanced the Moving Object Detection and Object Tracking for Traffic Surveillance Using RBF-FDLNN and CBF Algorithm," *Expert Systems with Applications*, Vol. 191, p. 116306, ISSN 0957-4174. 10.1016/j.eswa.2021.116306

[4] D. Liu, P. Ning, and R. Li. 2005. "Establishing Pairwise Keys in Distributed Sensor Networks," *ACM Transactions on Information Systems Security*, Vol. 8, No. 1, pp. 41–77.

[5] B. Lai, S. Kim, and I. Verbauwhede. 2002. "Scalable Session Key Construction Protocols for Wireless Sensor Networks." In Proceedings of the IEEE Workshop on Large Scale Real Time and Embedded Systems (LATES'02), 1–6, Austin, Texas, USA.

[6] S. Pandey, R. Miri, G. R. Sinha, and R. Raja. 2022. "AFD FiltEr and E2N2 Classifier for Improving Visualization of Crop Image and Crop Classification in Remote Sensing Image," *International Journal of Remote Sensing*, Vol. 43, No. 1, pp. 1–26. 10.1080/01431161.2021.2000062

[7] S. A. Cametepe and B. Yener. 2007. "Combinatorial Design of Key Distribution Mechanisms for Wireless Sensor Networks," *IEEE/ACM Transactions on Networksing (TON)*, Vol. 15, No. 2, pp. 346–358.

[8] J. Lee and D. R. Stinson. August 2004. "Deterministic Key Pre-Distribution Schemes for Distributed Sensor Networks." In Proceedings of the 11th International Workshop on Selected Areas in Crypography (SAC'04), 294–307, Waterloo, Canada, Selected Areas in Cryptography, pp. 294–307. Springer LNCS, Vol. 3357.

[9] R. Chandrakar, R. Raja, R. Miri, R. K. Patra, and U. Sinha, "2021. Computer Succored Vaticination of Multi-Object Detection and Histogram Enhancement in Low Vision," *Int. J. of Biometrics. Special Issue: Investigation of Robustness in Image Enhancement and Preprocessing Techniques for Biometrics and Computer Vision Applications*, Vol. 3, No.1, pp. 1–12.

[10] T. D. Diwan, S. Choubey, and H. S. Hota. 2021. "An Investigation and Analysis of Cyber Security Information Systems: Latest Trends and Future Suggestion," *International Journal IT in Industry*, Vol. 9, No. 2, pp. 477–492, ISSN: ISSN (Print): 2204-0595 ISSN (Online): 2203-1731.

[11] D. Liu and P. Ning. October 2003. "Location-Based Pair-Wise Key Establishments for Static Sensor Networks." In Proceedings of the ACM Workshop on Security in Ad hoc and Sensor Networks, pp. 72–82.

[12] Oladayo O. Olakanmi, A. Pamela, and A. Ashraf. 2018. "A Review on Secure Routing Protocols for Wireless Sensor Networks," *International Journal of Sensors, Wireless Communications and Control*. Vol. 7, No. 2, pp. 79–92.

[13] Tarun Dhar Diwan, Siddhartha Choubey, and H. S. Hota. entitled 2020. "Development of Real Time Automated Security System for Internet of Things (IoT)," *International Journal of Advanced Science and Technology*, Vol. 29, No. 6s, pp. 4180–4195, ISSN: 2005-4238.

[14] Oladayo O. Olakanmi and A. Dada. 2018. "An Efficient Point-to-Point Security Solution for Multi-hop Routing in Wireless Sensor Networks," *Security and Privacy*, Vol. 4, No. 5, pp. 73–84. 10.1002/spy2.58

[15] T. D. Diwan, S. Choubey, and H. S. Hota, "A Novel Hybrid Approach for Cyber Security in IoT Network Using Deep Learning Techniques," *International Journal of Advanced Science and Technology*, Vol. 29, No. 6S, pp. 4169–4179. ISSN:2394-5125, ISSN: 2005-4238.

[16] J. Xu and R. Li. 2006. "A New In-network Differentiated Services Mechanism in Wireless Sensor Networks." In 2006 International Conference on Communication Technology, pp. 1–3. doi: 10.1109/ICCT.2006.341969

[17] T. D. Diwan, S. Choubey, and H. S. Hota. 2021. "An Experimental Analysis of Security Vulnerabilities in Industrial Internet of Things Services," *International Journal of Information Technology in Industry IT in Industry*, Vol. 9, No. 3, pp. 592–612, ISSN: 2203-1731.

[18] K. Sohrabi, J. Gao, V. Ailawadhi, and G. J. Pottie. 2000. "Protocols for Self-Organization of a Wireless Sensor Network," *IEEE Personal Communications*, Vol. 7, No. Pt 5, pp. 16–27.

[19] A. Messaoudi, R. Elkamel, A. Helali, and R. Bouallegue. 2017. "Cross-Layer Based Routing Protocol for Wireless Sensor Networks Using a Fuzzy Logic Module." In Paper Presented at the 13th International Wireless Communications and Mobile Computing Conference (IWCMC).

[20] G. Murugaboopathi and Khaana. 2008. "Reliable Communications in Sensor Networks," *Journal of Engineering and Applied Science*, Vol. 3, pp. 911–917. Available from: https://medwelljournals.com/abstract/?doi=jeasci.2008.911.917

[21] J. Kumar, S. Tripathi, and R. K. Tiwari. 2016. "Routing Protocol for Wireless Sensor Networks Using Swarm Intelligence-ACO with ECPSOA." In International Conference of Information Technology.

[22] D. R. Huei-Wen. 2012. "A Secure Routing Protocol for Wireless Sensor Networks with Consideration of Energy Efficiency." In IEEE National Taiwan University of Science and Technology, pp. 224–232.

[23] H. Chan, A. Perrig, and D. Song. May 2003. "Random Key Pre-Distribution Schemes for Sensor Networks." In Proceedings of the IEEE Symposium on Security and Privacy (S&P'03), 197, Berkeley, California, USA.

[24] L. Tiwari, R. Raja, V. Awasthi, R. Miri, G. R. Sinha, and M. H. Alkinani. 2021. "Polat, Detection of Lung Nodule and Cancer Using Novel Mask-3 FCM and TWEDLNN Algorithms," *Measurement*, Vol. 172, p. 108882, ISSN 0263-2241. 10.1016/j.measurement.2020.108882

[25] W. Du, J. Deng, Y. S. Han, S. Chen, and P. K. Varshney. 2004. "A Key Management Scheme for Wireless Sensor Networks Using Deployment Knowledge." In Proceedings of IEEE INFOCOM, pp. 586–597, Hong Kong, China.

[26] S. Kumar, R. Raja, and A. Gandham. 2020. "Tracking an Object Using Traditional MS (Mean Shift) and CBWH MS (Mean Shift) Algorithm with Kalman Filter." In: P. Johri, J. Verma, S. Paul (eds.), *Applications of Machine Learning*. Algorithms for Intelligent Systems. Springer, Singapore. pp. 79–98. 10.1007/978-981-15-3357-0_4

[27] M. Amjad, M. K. Afzal, T. Umer, and B-S. Kim. 2017. "QoS-Aware and Heterogeneously Clustered Routing Protocol for Wireless Sensor Networks," *IEEE Access*. Vol. 5, pp. 10250–10262.

[28] R. Chandrakar, R. Raja, and R. Miri. 2021. "Animal Detection Based on Deep Convolutional Neural Networks with Genetic Segmentation," *Multimed Tools and Applications*, Vol. 73, No. 2, pp. 1–14. 10.1007/s11042-021-11290-4

[29] L. J. G. Villalba, A. L. S. Orozco, A. T. Cabrera, and C. J. B. Abbas. 2009. "Routing Protocols in Wireless Sensor Networks," *International Journal of Medical Sciences*, Vol. 9, No. 11, pp. 8399–8421.

[30] R. K. Lenka, A. K. Rath, Z. T., S. S, D. Puthal, N. V. R. Simha, and R. Raja. 2018. "Building Scalable Cyber-Physical-Social Networking Infrastructure Using IoT and Low Power Sensors," Vol. 6, No. 1, pp. 30162–30173, IEEE Access, Print ISSN: 2169-3536, Online ISSN: 2169-3536, Digital Object Identifier: 10.1109/ACCESS.2018.2842760

[31] R. Raja, T. S. Sinha, and R. P. Dubey. 2015. "Recognition of Human-Face from Side-View Using Progressive Switching Pattern and Soft-Computing Technique," *Association for the Advancement of Modelling and Simulation Techniques in Enterprises, Advance B*, Vol. 58, No. 1, pp. 14–34, ISSN: 1240-4543.

[32] R. Raja, R. k. Patra, and T. S. Sinha. 2017. "Extraction of Features from Dummy face for improving Biometrical Authentication of Human," *International Journal of Luminescence and Application*, Vol. 7, No. 3–4, October–December 2017, Article 259, pp. 507–512. ISSN:1 2277-6362.

12 An Epileptic Seizure Detection and Classification Based on Machine Learning Techniques

Lokesh Singh
Department of Computer Science and Engineering, Alliance University, Banglore, Karnataka, India

Rekh Ram Janghel
Department of Information Technology, National Institute of Technology, Raipur, Chhattisgarh, India

Satya Prakash Sahu
National Institute of Technology, Raipur, Chhattisgarh, India

CONTENTS

12.1	Introduction	170
12.2	Related Work	172
12.3	Proposed Methodology	173
	12.3.1 Database Description – BONN University EEG Dataset	174
	12.3.1.1 Data Pre-processing	176
	12.3.1.2 Statistical Features of the Dataset	176
	12.3.2 Evaluation Assessment Method	178
	12.3.3 Classification Techniques	178
	12.3.3.1 Support Vector Machines (SVMs)	178
	12.3.3.2 Random Forests	179
	12.3.3.3 Extreme Learning Machine	179
	12.3.3.4 K-Nearest Neighbors	179
	12.3.3.5 Logistic Regression	180
	12.3.3.6 Decision Trees	180
	12.3.3.7 Multilayer Perceptron	180
	12.3.3.8 Ensemble Classifiers	180
12.4	Experimental Results	181

DOI: 10.1201/9781003217091-12

12.5 Discussion...182
12.6 Conclusion..182
References..183

12.1 INTRODUCTION

There are a range of common neurological diseases referred to as epilepsy, and the term is used to describe seizures that are uncontrollable due to abnormal electrical discharges in the brain. According to a World Health Organization (WHO) estimate, epilepsy affects more than 50 million people worldwide. One in every 100 people may experience a spasm at some point in their lives, and epilepsy affects an estimated 2.4 million people each year. Epilepsy's symptoms and treatment options were unknown until recently [1]. It is possible for epileptic seizures to create a wide range of unpleasant bodily, social, and mental effects. These include loss of consciousness, harm, and even death (Table 12.1).

Electroencephalogram (EEG) is a low-density sign generated in the brain as a result of a data stream generated by a few neurons communicating with each other. A comprehensive examination of these indicators could help us understand a wide range of human brain problems [2]. The data stream is handled by a large number of neurons in the human brain. Because of this data progression, the human body behaves similarly. Our brains contain billions of neurons. A low-frequency electric charge is created when a neuron comes into contact with another neuron [3]. As a result, several systems exist to deal with this type of mechanism.

TABLE 12.1
Different waveforms present in the brain

Waveform	Range (Hz)	Preferred Locality	Symptoms
Delta	Less than 4	Frontally in adults, posteriorly in children, high amplitude waves	During some continuous-attention tasks, adult slow-wave sleep has been discovered.
Theta	4 to 7	Found in places unrelated to the task at hand	Drowsiness is more common in young children than in adults and teenagers.
Alpha	8 to 15	On dominant mode, the posterior areas of the skull have a larger amplitude on both sides	Relaxed/reflective behaviour, including closing the eyes, is linked to inhibition control and inhibitory activity in many parts of the brain.
Beta	16 to 31	Waves of low amplitude on both sides, symmetrical distribution, most visible frontally	Active thinking, concentration, heightened vigilance, and apprehension.
Gamma	Greater than 32	Somatosensory cortex	During cross-modal sensory processing, the display is shown when matching recognised objects in short-term memory.

An Epileptic Seizure Detection

Epilepsy commences from the Greek note 'Epilepsia,' which translates as 'seizure' or 'to take advantage of.' It is an actual neurological problem with noticeable qualities, tending to intermissive seizures [1]. More than 3000 years ago, epilepsy was established in the Babylonian content on medication. This affliction isn't restricted to individuals, but stretches out to cover all types of well-evolved creatures like canines, felines, and rodents [2]. Be that as it may, the word "epilepsy" doesn't give any signs about the reason or seriousness of the seizures; it is unexceptional and consistently circulated throughout the planet [3]. A few speculations about the reason are now accessible. The fundamental driver is electrical action aggravation inside a cerebrum, which could begin for a few reasons like deformities, lack of oxygen at the time of labor, and low blood sugar. Universally, epilepsy impacts around 50 million people, where, once in their life in any event, 100 million have been affected. In most cases, it represents 1% of the world's disease weight, with a pervasiveness rate of 0.5%–1%. More than one seizure by a patient is a cardinal adverse effect of epilepsy [4]. It was used to explain a quick dissolution or dramatic action in the cerebrum that forced a patient's observance, horripilation, and wound of transient cognizance to change. Perpetually, seizures of a couple of minutes(s) can occur when there is absense of air. This prompts genuine wounds, including broken bones and now and again death [5].

In light of the manifestations, seizures are arranged by neuro-specialists into two principal classes, fractional and summed up—as demonstrated in Figure 12.1. Incomplete seizure, likewise called 'central seizure,' means just a part of the cerebral half of the globe is influenced. There are two sorts of partial seizures: basic halfway and complex-fractional. A patient doesn't pass out in the basic fractional yet can't convey as expected [6]. In the perplexing fractional, an individual gets confused about the environment and starts behaving unusually like biting and murmuring; this is known as 'central weakened mindfulness seizure.' In actuality, in complete seizures, all districts of the cerebrum suffer and whole mind networks get influenced rapidly [7]. Summed up, seizures are of numerous kinds; however, they are extensively separated into two classes: convulsive and non-convulsive.

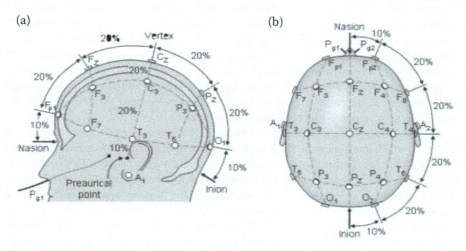

FIGURE 12.1 Standardized electrode placement scheme.

Uses of artificial intelligence (AI) are essentially seen on well-being and natural informational collections for better results. Analysts/researchers on various territories, explicitly information mining and AI, effectively propose answers for better seizure location. AI has been essentially applied to find practical and significant examples from various area datasets [8,9]. It assumes a critical and potential part in tackling the issues of different orders like medical services. On cerebrum datasets, AI was also used for seizure localization, epilepsy lateralization, differentiating seizure states, and limiting. ANN, SVM, choice tree, choice woodland, and arbitrary backwoods are some of the AI classifiers that have completed this task. Unquestionably, before various audits have been completed on seizure identification alongside applied highlights, classifiers, and asserted exactness without zeroing in on the difficulties, researchers looked at the information while investigating datasets of neurological problems [10].

12.2 RELATED WORK

Researchers have come up with a variety of methods for calculating time-based signals. Many applications have been built using the method proposed by these scholars. They outperform all other assessment methods due to the importance of these signals, which are also referred to as wavelet signals. Up to 90% accuracy was achieved using discrete wavelet transformation techniques to divide the waves into spike, sharp, and measuring spike waves, all of which are seizure class waves, as described by Adeli et al. [11].

Many procedures based on the theory of sign examination have been applied to obtain portrayals and concentrate the highlights of enthusiasm for characterization goals. EEG research utilised neural systems and factual example acknowledgement techniques. There have been numerous frameworks for neural network (NN) recognition presented by analysts. There was a large false discovery rate in Pradhan et al. [12], whereas Weng and Khorasani [13] used the highlights proposed by Gotman with a variable neural system framework. Intermittent neural networks with wavelet pre-processing, as demonstrated by Petrosian et al. [14], can anticipate the beginning of seizures and detect epilepsy even on scalp and cerebral EEGs with only one channel. Bioelectrical information, specifically neural accounts and 128-channel EEG, was evolved by Folkers et al. [15] into a flexible sign preparation and evaluation approach to enable faster and more accurate estimates. Wavelet change computations done continuously on a current advanced sign processor stage degrade the sign into sub-groups within this structure.

For the most part, the difficulties with two-class seizure detection are due to the order in which conventional EEG sections taken from healthy people (set A) and seizure EEG patterns from epileptic patients having dynamic seizures (set E) are collected [16]. Robotic seizure detection frameworks have been offered that include many EEG highlights retrieved from time, recurrence space, wavelet area, as well as auto-backward coefficients and cepstral highlights [17,18]. Two hidden layers of the back-engendering neural organisation (BNN) classifier used these features to get a normal order accuracy of 93.000%. Using wavelet change to establish EEG recurrence groups,

An Epileptic Seizure Detection

Subasi was able to achieve a normal order precision of 94.50% by incorporating each of the alien segments into the combination of specialists (ME) classifiers. Polat et al. used a decision tree (DT) classifier to obtain 98.68% characterisation accuracy. In addition, the EEG cross-relationship coefficients were employed to analyse three facts and then submit them to the SVM for EEG order as an element vector. This model's seizure detection accuracy was 95.96%. It was possible to attain approximately comparable location accuracy using the extreme learning machine (ELM) classifier with several non-direct highlights, such as estimated entropy and Hurst type [19]. Wavelet change was used to separate the EEG signals into five estimation and detail sub-groups. To disperse EEG signals into five distinct recurrence rhythms, the wavelet change was employed in [20]. These beats were subjected to an SVM classifier in order to remove any factual or non-straight highlights that might skew the accuracy of the recognition. According to [21], using the SVM and weighted stage entropy, Song et al. obtained an order exactness of 97.25%. To further split down the EEG signals into various sub-groups, the staggered wavelet change was used in [22]. The alien highlights were then deleted, and the element vector was created. As a result, the ELM for preparation and order was familiarised with the component vector, which resulted in a promising 99.48% affectability. For the recognition of seizure scenes, Guoet al. used wavelet-based entropy approximation (WBE) in conjunction with an ANN model that had a normal order accuracy of 98.27%. KNN classifier was used in conjunction with a genetic computation created by [23] to reach an accuracy of 98.40% when determining automated EEG highlights.

12.3 PROPOSED METHODOLOGY

In this section, we show a visual representation of the model used to recognize seizures from an EEG/ECoG seizure dataset provided by BONN University [24], as shown in Figure 12.2. Data collection, data preparation, and data pre-handling;

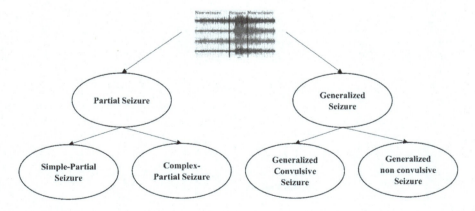

FIGURE 12.2 Types of seizure and its subtypes.

features extraction using the PyEEG module; machine learning classifiers and neural network classifiers; and performance evaluation are the four steps of the interaction.

12.3.1 Database Description – BONN University EEG Dataset

The dataset of thought signals must be gathered as a starting point. Various observing devices are used for this [25]. EEG and ECoG are the most commonly utilized devices because their channels or anodes are embedded on the outside of the scalp using a stick according to the 10–20 International framework at various projections. Each has a wire connection to the EEG device, which provides relevant information regarding voltage variations as well as worldly and spatial data [26]. As shown in Figure 12.3, the EEG channels are placed on the subject's scalp, and the EEG analyzing apparatus examines the electrical signals, displaying them on the screen. Furthermore, the investigator thoroughly examined these simple symptoms and classified them as 'seizure' or 'non-seizure' states [23].

This reveals the fundamental steps for gathering the dataset via EEG medium, displaying crude EEG signals, converting EEG signs to a two-dimensional table, highlight selection, setting up the dataset with seizure (S) and non-seizure (NS), applying AI classifier(s) and seizure recognition, and other related tasks (Figure 12.4).

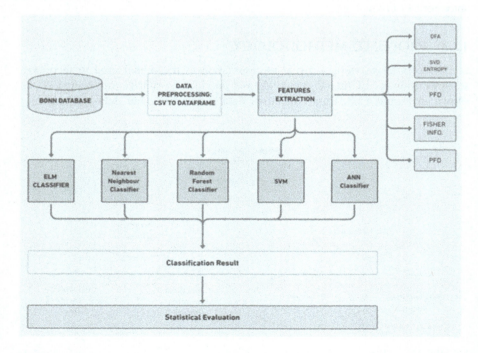

FIGURE 12.3 Block diagram.

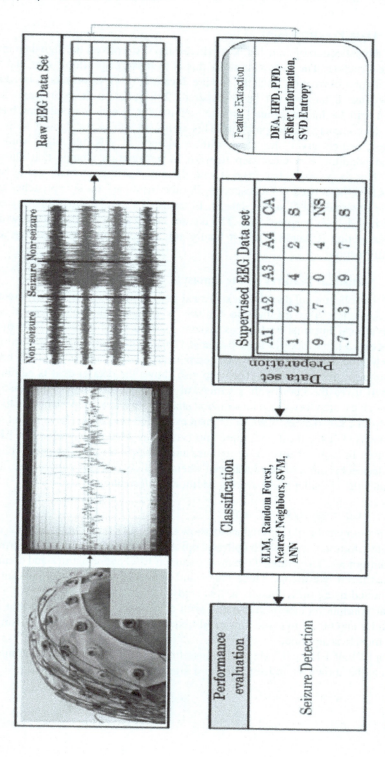

FIGURE 12.4 Basic model of epileptic seizure detection.

12.3.1.1 Data Pre-processing

Following the collection of data, the next critical step is to convert the sign data into a 2-D table structure. The justification for this is to make inspection easier and to provide critical information such as seizure location. This information is rudimentary because it has not yet been prepared [27]. As a result, providing vital information will be inappropriate. Various component selection techniques have been used to complete the preparation. This step also shows the dataset as it is administered, which gives the class with a variety of possible class values. Information handling is a clear step forward in information change that allows important data to be extracted from a raw dataset. Various element extraction procedures have been used in this capacity. For the most part, these approaches are used on the separated EEG signal dataset. In terms of several factual measure esteems, the raw dataset becomes rich. The dataset becomes more educational when highlight extraction is handled, and it finally aids the classifier in recovering additional information.

12.3.1.2 Statistical Features of the Dataset

Following the collection of data, the next critical step is to convert the sign data into a 2-D table structure. The justification for this is to make inspection easier and to provide critical information such as seizure location. This information is rudimentary because it has not yet been prepared [20]. As a result, providing vital information will be inappropriate. Various component selection techniques have been used to complete the preparation. This step also shows the dataset as it is administered, which gives the class with a variety of possible class values. Information handling is a clear step forward in information change that allows important data to be extracted from a raw dataset. Various element extraction procedures have been used in this capacity. For the most part, these approaches are used on the separated EEG signal dataset. In terms of several factual measure esteems, the raw dataset becomes rich. The dataset becomes more educational when highlight extraction is handled, and it finally aids the classifier in recovering additional information.

12.3.1.2.1 Common EEG Artifacts

It's not uncommon for old anomalies to show up in EEG recordings. Seizure designs and the accuracy with which epileptic episodes are discovered may be harmed by these antiquities. They studied the most common EEG abnormalities and created models that acted like them." Three major and unavoidable sources of antiquities were examined using these models in this study:

Regular commotion can be illustrated with a bandpass channel (BPF) that separates 20 Hz and 60 Hz to avoid unwanted frequencies, and then replicated using a common muscle scalp map.

Using a BPF of 1 Hz and 3 Hz to eliminate frequencies, the blinks and squints of the eyes can be shown as an irregular noise signal.

An Epileptic Seizure Detection 177

White Noise: A white Gaussian noise with greater substance depicts the electrical and environmental commotion. After adding muscle antiques, eye-flickering, and repeating sound, Figures 12.1b, c, and d depict the contaminated variants of a similar indication. As depicted in Figures 12.1e, f, g, and h, the recurrence spectra found in Figures 12.1a, b, c, and h are depicted in these Figures (d). Violent EEG signals are generated by muscle spasm, eye flashing, and reverberating sounds that have been trained to produce different sign-to-commotion ratios (SNRs). In order for the commotion signal to have the same strength as the EEG signal, its SNR is set to 0 dB.

12.3.1.2.2 Features Extraction

PyEEG's target customers are developers working in the field of computational neuroscience. PyEEG is a Python package that focuses solely on extracting highlights from EEG/MEG data. PyEEG makes use of Python's standard library, SciPy, and an acceptable Python module to provide understandable input. PyEEG does not employ any novel data structures and instead uses conventional Python and NumPy formats to show information. The consequence is that we need to improve our usage of PyEEG, particularly for customers who aren't on any sort of planning schedule. The time to collect periods as a series of dense numbers of dense points and multiple awareness barriers is when all powers are dedicated. Borders have default ratings. The output function's output is the flow point number if the object is a scale or a rundown of the vector numbers, however. In our proposed process, we used five key points separated from the EEG database using the PyEEG module to specify detrended fluctuation analysis, Higuchi fractal dimension, single value decoration, Fisheries information, and Petrosian fractal dimension (Figure 12.5).

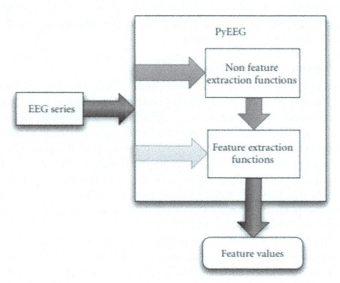

FIGURE 12.5 PyEEG framework.

12.3.2 EVALUATION ASSESSMENT METHOD

TABLE 12.2
Performance evaluation measures

Metric Name	Mathematical Formula	Description
Accuracy	$\frac{TP + TN}{TP + TN + EP + FN}$	The degree to which a value's projected or categorised value comes near to its actual value is known as accuracy. True accuracy is a term for the state of being accurate. It is the easiest to understand metric.
Sensitivity/ Recall	$\frac{TP}{TP + FN} \times 100$	Sensitivity is the ratio of the correctly +ve labeled by our algorithm to all those who have seizure in reality.
Specificity	$\frac{TN}{TN + FP} \times 100$	Specificity measures how effectively a classifier measures the negative labels.
Precision	$\frac{TP}{FP + FP} \times 100$	It is a class agreement between the data labels and the positive labels the classifier gives.
F1 Score	2 * (Recall * Precision)/ (Recall + Precision)	Both precision and recall are considered for calculating the F1 score. For this, average of the precision and recall (harmonic mean) is calculated.

12.3.3 CLASSIFICATION TECHNIQUES

Arrangement strategy in artificial intelligence and measurements is an administered learning approach. The PC program gains from the information given to it and mentions new objective facts or orders.

12.3.3.1 Support Vector Machines (SVMs)

As a solid double classifier, SVM provides high-order precision in a variety of applications. The goal of the assistance vector machine approach is to find a hyperplane in an N-dimensional space that arranges information concentrates extraordinarily well [12]. We can choose from a variety of possible hyperplanes to partition any two types of information focuses. Our goal is to find a plane with the greatest severe edge. Increasing the edge distance provides some support, allowing subsequent information focuses to be more confidently grouped. The disadvantage of help vector machines [sir 1, sir 32] is that they are limited to just grouping two-fold concerns. The SVM calculation's expense (C) boundary was set to 1, which is the SVM calculation's default worth for all grouping measures [12].

The essential ideas of the VC hypothesis and the risk minimization method underpin support vector. The hyperplane is derived in this machine learning classifier, which is separated into several classes. Researchers have presented the support vector classifier, which, unlike other machine learning classifiers, estimates the best value of the performance indicators at the time. These are the vectors that are called the support vectors [28,29].

12.3.3.2 Random Forests

Relapse, planning, and other jobs can all benefit from the use of random forests. The arbitrary timberland classifier performs better than a single individual tree since it is a group classifier with multiple trees [25]. It is possible to utilise a random Timberland classifier to do both relapse and grouping. Random forest receives a class vote in favour of the huge number of trees when employed in collecting errands, and the arrangement is then dependent on the majority vote [30,31]. Aggregating (voting in favour of or averaging for relapse) the gathering's projections creates the expectation: Assume that there are N occurrences in your training set, but that the original data has been replaced by N random selections from your original data. This sample will serve as a training set for the tree's development. At each node, we choose m random variables from the M and utilise the best split based on this m to divide the input data into its component parts, with each node specifying mM as an input parameter. m's value does not change as the forest grows. Clipped to the maximum extent possible, each tree has been pruned. In our trial, no trimming is done.

An ensemble classifier called a "random forest classifier" uses decision trees to determine the metrics of a given dataset or value. Because it is a part of an ensemble classifier, this classifier has excellent accuracy, can be used on huge datasets, and offers methods for stabilising error that aid in model improvement. It can also be used on unsupervised data, which aids in maintaining unbalanced data.

12.3.3.3 Extreme Learning Machine

Extreme learning machines are feed-forward neural organizations for grouping, relapse, bunching, meager guess, pressure, and highlight learning with a solitary layer or different layers of covered- up hubs, where the boundaries of covered-up hubs (not simply the load-associating contributions to covered-up hubs) need not be tuned [26]. These secret hubs can be arbitrarily appointed and never refreshed (for example, random projections but with nonlinear changes) or acquired from their precursors without being changed. By and large, the yield loads of covered-up hubs are generally scholarly in a solitary advance, which sums to learning a straight model. The name "extreme learning machine" (ELM) was given to such models by its principal creator Guang-Bin Huang [27].

12.3.3.4 K-Nearest Neighbors

In light of the KNN arrangement's reasoning technique, we expect the test archive to bear the same name as preparation reports that are located in the vicinity of it. There are Voronoi cells when a large number of items are Veronoi ornamented, and each item's phone is made up of all signals closer to it than to other things. In terms of depiction and backsliding, the K-nearest neighbour (KNN) method is a non-parametric alternative.

Using KNN in machine learning may be the most fundamental order computation in the field. Design acknowledgment, information mining, and disruption discovery are all areas where it has a position in the regulated learning environment. For both characterisation and relapse issues, the KNN calculation [32] is an easy-to-carry-out AI calculation that can be used. The KNN computation assumes that similar things are located close to one another. Relatively speaking, most objects are in close proximity. KNN utilises some science to catch the probability of similitude (also known as distance, proximity, or closeness) as it is described below in KNN calculation [33]:

The first step is to load the training data and set the value of K. Identify all records in the training set that have a Euclidean distance of more than 2. After sorting, add the spaces to your sorted list and pick the K-th item from the top. Classify a test point according to how many similarities there are between it and one or more of the reference points.

12.3.3.5 Logistic Regression

There are various mode extensions since it is a form of statistical model that uses a logistic function to model a binary dependent variable. The formula for logistic regression is:

$$\text{Logit}(P1) = \log(p1/1 - p1) = \beta 0 + \beta 1 X1 + \ldots\ldots + \beta n Xn \quad (12.1)$$

$$= \beta 0 + \Sigma \beta i\ Xi \quad (12.2)$$

In Eq. (12.1), $\beta 0$ is the intercept and $\beta 1$, $\beta 2 \ldots \beta n$ are the coefficients combined with the analytical variable $X1$, $X2 \ldots Xn$ [34]. One of the most essential characteristics of this classifier is that it works with a large dataset, has high accuracy because it is part of a various classifiers, and provides ways for managing error that aid in model improvement.

12.3.3.6 Decision Trees

A decision tree is a form of supplementary tool that use a tree-like structure to represent potential results, such as probability, resourced costs, and utility. This type of technique is used to express conditional statements. The test on the property is represented by each internal node in the tree-like structure. Each branch that is present in the tree indicates a test result, and each leaf node that is present represents a class label [35,36].

12.3.3.7 Multilayer Perceptron

A neural network with multiple layers is known as a multilayer perceptron. Weighted multilayer perceptrons are used to calculate the results of a signal that is passed through an input layer. In many ways, it's like a perceptron network, but instead of having only one layer, it has several. Based on the weights in the neural network, it generates the best possible results. In the hierarchy, some researchers believe this neural network outperforms other networks [37,38].

12.3.3.8 Ensemble Classifiers

Ensemble classifiers are classifiers in which numerous classifiers are merged to form a single classifier in order to increase the algorithm's performance. Each split in the training and testing data can yield an infinite number of increments. The best results are produced by a supervised neural network, in which the weights of the neural network are used to optimise the network. Several researchers have found that this neural network performs better than others in the hierarchy [39,40]. For a better outcome estimate, we combined the bagging classifier and AdaBoost classifier with the decision tree classifier and random forest classifier.

12.4 EXPERIMENTAL RESULTS

Jupyter notebooks with Scikit-learn modules were used to implement all of the models. Hosted on Github, the dataset can be accessed remotely and processed more quickly, run through 100 iterations (Tables 12.3 and 12.4) (Figure 12.6).

TABLE 12.3
Results of EEG signal dataset using machine learning models

Sl. No.	Model/Technique	Acc	Spec	Sen	F1	Prec
1.	ELM Neural Network (ELM)	99.93	94.34	93.23	98.12	98.87
2.	K-Nearest Neighbors (KNN)	95.87	90.52	92.32	93.63	94.12
3.	SVM (Gaussian, Sigmoid)	85.85	81.67	79.54	84.20	83.57
4.	Random Forest (RF)	98.84	96.12	94.93	98.28	98.89
5.	ANN	68.27	65.12	63.82	65.67	65.42
6.	Logistic Regression (LR)	82.64	78.43	77.65	80.36	81.96
7.	Decision Tree (DT)	85.75	81.24	79.32	83.95	84.57
8.	Multi-Layer Perceptron (MLP)	92.56	89.67	85.47	91.27	92.35
9.	Naïve Bayes (NB)	82.64	80.31	78.41	81.85	82.37
10.	Bagging Classifier (BC)	94.36	90.48	88.37	93.51	94.25
11.	Ada Boost Classifier (ABC)	95.41	91.36	89.84	94.24	94.69

Acc – Accuracy, Spec – Specificity, Sen – Sensitivity, F1 – F1Score, Prec – Precision.

TABLE 12.4
A comparative study of proposed work with the existing work for the classification

Sl. No.	References	Features Extraction Methods	Classifier	Overall Accuracy
1.	Weng [15]	Dominant frequency and mean power spectrum	Linear SVM	98
2.	Rashid et al. [12]	Discrete wavelet transform (DWT)	BNN	93
3.	Kumar et al. [40]	–	LDA	97.52
4.	Aarabi et al. [39]	Mean and variance	BLDA	96.67
5.	Proposed Method	Five features using PyEEG	ELM	99.93

182 Next Generation Healthcare Systems Using Soft Computing Techniques

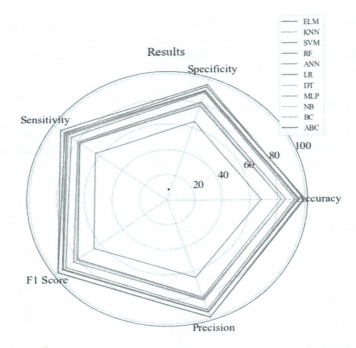

FIGURE 12.6 Performance comparison of machine learning methods on different measures.

12.5 DISCUSSION

Epilepsy may be the most common neurological condition. Described as a seizure, it is the second most common neurological problem after a stroke, according to the WHO [9]. Fainting is possible without paying attention to certain situations. Patients with epilepsy experience the unintended consequences of sudden unconsciousness. They cannot rely on themselves and cannot protect themselves from death or injury due to passing out and car crashes. To date, these conditions are still treated with medication and medical procedures; no correction is available, and treatments containing anticonvulsants are not fully robust for all forms of epilepsy. EEG plays a vital role in the diagnosis of epilepsy, as it measures the comparison of electrical energy exchange between the anodes along the scalp by the ionic flow of cerebrum neurons and provides transient and spatial information about the brain. EEG detection requires close medical examination as an open measure of time and effort [41,42].

12.6 CONCLUSION

This chapter presented a productive ELM classifier dependent on a couple of straightforward highlights for epileptic seizure discovery in EEG signals. We examined the presentation of highlights independently and on the whole in differing size investigation windows. We saw that by utilizing five highlights separated using the PyEEG module, the classifier delivered ideal outcomes.

An Epileptic Seizure Detection

We noticed that the length of the investigation window assumes an essential part in arrangement execution. Further examination is justified to investigate this intricacy versus execution compromises altogether.

REFERENCES

[1] S. K. Satapathy, S. Dehuri, and A. K. Jagadev, "EEG signal classification using PSO trained RBF neural network for epilepsy identification," *Informatics Med. Unlocked*, vol. 6, no. June 2016, pp. 1–11, 2017.

[2] P. P. Muhammed Shanir, Yusuf U. khan, and Omar Farooq, "Time domain analysis of EEG for automatic Seizure detection," *Emerging Trends in Electrical and Electronics Engineering*, vol. 1, pp. 1–12, 2015.

[3] U. J. Chaudhary, J. S. Duncan, and L. Lemieux, "A dialogue with historical concepts of epilepsy from the Babylonians to hughlings jackson: Ppersis-tent beliefs," *Epilepsy Behav*, vol. 21, no. 2, pp. 109–114, 2011.

[4] L. Tiwari, R. Raja, V. Awasthi, R. Miri, G. R. Sinha, and M. H. Alkinani, "Polat, detection of lung nodule and cancer using novel Mask-3 FCM and TWEDLNN algorithms," *Measurement*, vol. 172, p. 108882, 2021, ISSN 0263-2241. 10.1016/j.measurement.2020.108882

[5] R. Chandrakar, R. Raja, R. Miri, U. Sinha, A. K. S. Kushwaha, and H. Raja, "Enhanced the moving object detection and object tracking for traffic surveillance using RBF-FDLNN and CBF algorithm," *Expert Systems with Applications*, vol. 191, p. 116306, 2022. ISSN 0957-4174. 10.1016/j.eswa.2021.116306

[6] H. Qu and J. Gotman, "Improvement in seizure detection performance by automatic adaptation to the eeg of each patient," *Electroencepha- logr Clin Neurophysiol*, vol. 86, no. 2, pp. 79–87, 1993.

[7] R. Raja, S. Kumar, and Md Rashid, "Color object detection based image retrieval using ROI segmentation with multi-feature method," *Wireless Personal Communication Springer Journal*, Print ISSN0929-6212 online ISSN1572-834, vol. 112, no. 1, pp. 169–192, 2020. 10.1007/s11277-019-07021-6

[8] S. Schachter, P. Shafer, and J. Sirven, *What Causes Epilepsy and Seizures*. Epilepsy Foundation. 2013.

[9] W.. Zhou, Y. Liu, Q. Yuan, and X. Li, "Epileptic seizure detection using lacunarity and bayesian linear discriminant analysis in intracranial EEG," *IEEE Transactions on Biomedical Engineering*, vol. 60, no. 12, pp. 3375–3381, 2013.

[10] S. Pandey, R. Miri, G. R. Sinha, and R. Raja, "AFD filter and E2N2 classifier for improving visualization of crop image and crop classification in remote sensing image," *International Journal of Remote Sensing*, vol 43, no. 1, pp. 1–26, 2022. 10.1080/01431161.2021.2000062

[11] WHO: Media Center Epilepsy, (Fact sheet N999). http://www.who.int/mediacentre/factsheets/fs999/en/ (2015) Accessed 15 July 2019.

[12] R. Raja, R. k. Patra, and T. S. Sinha, "Extraction of Features from Dummy face for improving Biometrical Authentication of Human," *International Journal of Luminescence and Application*, ISSN:1 2277-6362, vol. 7, no. 3–4, October–December 2017, Article 259, pp. 507–512, 2017.

[13] H. Adeli, Z. Zhou, and N. Dadmehr, "Analysis of EEG records in an epileptic patient using wavelet transform," *J. Neurosci. Methods*, vol. 123, no. 1, pp. 69–87, 2003.

[14] N. Pradhan, P. K. Sadasivan, and G. R. Arunodaya, "Detection of seizure activity in EEG by an artificial neural network: A preliminary study," *Comput. Biomed. Res.*, vol. 29, no. 4, pp. 303–313, 1996.

[15] W. Weng and K. Khorasani, "An adaptive structure neural networks with application to EEG automatic seizure detection," *Neural Networks*, vol. 9, no. 7, pp. 1223–1240, 1996.

[16] A. Petrosian, D. Prokhorov, R. Homan, R. Dasheiff, and D. Wunsch, "Recurrent neural network based prediction of epileptic seizures in intra- and extracranial EEG," *Neurocomputing*, vol. 30, no. 1–4, pp. 201–218, 2000.

[17] A. Folkers, F. Mösch, T. Malina, and U. G. Hofmann, "Realtime bioelectrical data acquisition and processing from 128 channels utilizing the wavelet-transformation," *Neurocomputing*, vol. 52–54, pp. 247–254, 2003.

[18] R. Chandrakar, R. Raja, R. Miri, R. K. Patra, and U. Sinha, "Computer succored vaticination of multi-object detection and histogram enhancement in low vision," *Int. J. of Biometrics. Special Issue: Investigation of Robustness in Image Enhancement and Preprocessing Techniques for Biometrics and Computer Vision Applications*, vol. 3, no. 1, pp. 1–25, 2022.

[19] R. S. Fisher, "The new classification of seizures by the inter-national league against epilepsy 2017," *Curr Neurol Neurosci Rep*, vol. 17, no. 6, p. 48, 2017.

[20] R. Raja, T. S. Sinha, and R. P. Dubey, "Recognition of human-face from side-view using progressive switching pattern and soft-computing technique," *Association for the Advancement of Modelling and Simulation Techniques in Enterprises, Advance B*, vol. 58, no. 1, pp. 14–34, 2015. ISSN: 1240-4543.

[21] M. Mahmud, M. S. Kaiser, A. Hussain, and S. Vassanelli, "Applications of deep learning and reinforcement learning to biological data," *IEEE Trans Neural Networks Learn Syst*, vol. 29, no. 6, pp. 2063–2079, 2018.

[22] Fayyad U. M., Piatetsky-Shapiro G., Smyth P., Uthurusamy R. (eds.), *Advances in Knowledge Discovery and Data Mining*. American Associa-tion for Artificial Intelligence, Menlo Park, CA. 1996.

[23] A. H. Ansari et al., "Weighted Performance Metrics for Automatic Neonatal Seizure Detection Using Multiscored EEG Data," *IEEE J. Biomed. Heal. Informatics*, vol. 22, no. 4, pp. 1114–1123, 2018.

[24] G. A. Singh and P. K. Gupta, "Performance analysis of various machine learning-based approaches for detection and classification of lung cancer in humans," *Neural Comput Appl*, vol. 31, no. 10, pp. 6863–6877, 2019.

[25] R. G. Andrzejak, G. Widman, K. Lehnertz, C. Rieke, P. David, and C. E. Elger, "The epileptic process as nonlinear deterministic dynamics in a stochastic environment: an evaluation on mesial temporal lobe epilepsy," *Epilepsy Research*, vol. 44, pp. 129–140, 2001.

[26] R. Chandrakar, R. Raja, and R. Miri, "Animal detection based on deep convolutional neural networks with genetic segmentation," *Multimed Tools and Applications*, vol. 162, no. 1, 2021. 10.1007/s11042-021-11290-4

[27] A. K. Tiwari, R. B. Pachori, V. Kanhangad, and B. K. Panigrahi, "Automated diagnosis of epilepsy using key-point-based local binary pattern of EEG signals," *IEEE J. Biomed. Heal. Informatics*, vol. 21, no. 4, pp. 888–896, 2017.

[28] T. Inouye, K. Shinosaki, H. Sakamoto et al., "Quantification of EEG irregularity by use of the entropy of the power spectrum," *Electroencephalography and Clinical Neurophysiology*, vol. 79, no. 3, pp. 204–210, 1991.

[29] S. J. Roberts, W. Penny, and I. Rezek, "Temporal and spatial complexity measures for electroencephalogram based brain- computer interfacing," *Medical and Biological Engineering and Computing*, vol. 37, no. 1, pp. 93–98, 1999.

[30] C-C. Chang and C-J. Lin, "LIBSVM: A library for support vector machines," *ACM Transactions on Intelligent Systems and Technology*, vol. 2, no. I3, April 2011.

[31] L. Breiman, "Bagging predictors," *Machine Learning*, vol. 26, no. 2, pp. 123–140, 1996.

[32] Extreme Learning Machine: https://towardsdatascience.com/build-an-extreme-learning-machine-in-python-91d1e895859
[33] S. Kumar, A. Jain, A. P. Shukla, S. Singh, R. Raja, and S. Rani, "A comparative analysis of machine learning algorithms for detection of organic and nonorganic cotton diseases," *Mathematical Problems in Engineering*, vol. 2021, Article ID 1790171, 18 pages, 2021. 10.1155/2021/1790171
[34] M. Kaczorowska, M. Plechawska-Wojcik, M. Tokovarov, and R. Dmytruk, "Comparison of the ICA and PCA methods in correction of EEG signal artefacts," in 2017 10th Int. Symp. Adv. Top. Electr. Eng. ATEE 2017, pp. 262–267, 2017.
[35] R. M. Aileni, S. Paşca, and A. Florescu, "Epileptic seizure classification based on supervised learning models," in 2019 11th Int. Symp. Adv. Top. Electr. Eng. ATEE 2019, no. c, pp. 1–4, 2019.
[36] Kalbkhani and M. G. Shayesteh, "Stockwell transform for epileptic seizure detection from EEG signals," *Biomed. Signal Process. Control*, vol. 38, pp. 108–118, 2017.
[37] E. Abdulhay, V. Elamaran, M. Chandrasekar, V. S. Balaji, and K. Narasimhan, "Automated diagnosis of epilepsy from EEG signals using ensemble learning approach," *Pattern Recognit. Lett.*, pp. 1–8, 2017.
[38] J. R. Quinlan, *C4.5: Programs for Machine Learning*. Morgan Kauf-Mann Publishers Inc., San Francisco. 1993.
[39] F. W. Aarabi, and R. Grebe, "Automated neonatal seizure detection: A multistage classification system through feature selection based on relevance and redundancy analysis," *Clinical Neurophysiology*, vol. 117, no. 2, pp. 328–340, 2006.
[40] Kumar and M. H. Kolekar, "Machine learning approach for epileptic seizure detection using wavelet analysis of EEG signals," in Medical Imaging, M-Health and Emerging Communication Systems (MedCom), 2014 International Conference. IEEE, pp. 412–416, 2014.
[41] A. K. Sahu, S. Sharma, and M. Tanveer, "Internet of Things attack detection using hybrid deep learning model," *Computer Communications*, vol. 176, pp. 146–154, 2021. ISSN 0140-3664. 10.1016/j.comcom.2021.05.024
[42] R. Raja, S. Kumar, S. Choudhary, and H. Dalmia, "An effective contour detection based image retrieval using multi-fusion method and neural network," *Submitted to Wireless Personal Communication, PREPRINT (Version 2)* available at Research Square. 2021. 10.21203/rs.3.rs-458104/v1

13 Analysis of Coronary Artery Disease Using Various Machine Learning Techniques

Saroj Kumar Pandey
Department of Computer Engineering and Applications, GLA University, Mathura (U.P.), India

Rekh Ram Janghel
Department of Information Technology, National Institute of Technology, Raipur, Chhattisgarh, India

Shubham Shukla
Department of Electronics & Communication Engineering, Krishna Institute of Engineering & Technology-Ghaziabad, Ghaziabad (U.P.), India

Yogadhar Pandey
Department of Computer Science & Engineering, Technocrats Institute of Technology-Excellence, Bhopal (M.P.), India

CONTENTS

13.1 Introduction 188
13.2 Literature Survey 190
13.3 Material and Methods 192
    13.3.1 Dataset Description 192
13.4 Methodology 192
    13.4.1 Data Normalization 192
        13.4.1.1 Data Splitting 193
        13.4.1.2 Classification Models 193
        13.4.1.3 Support Vector Machine 193
        13.4.1.4 Decision Tree 194
        13.4.1.5 Random Forest 194
        13.4.1.6 K-Nearest Neighbor (K-NN) 195
    13.4.2 Logistic Regression 195

13.4.3 Types of Logistic Regression 195
    13.4.3.1 Logistic Regression Assumptions 195
    13.4.3.2 Naïve Bayes 196
    13.4.3.3 XG-Boost 196
13.5 Result Analysis 196
    13.5.1 Performance Comparison of Algorithms 197
13.6 Discussion 199
13.7 Conclusion 204
References 205

13.1 INTRODUCTION

The set of conditions affecting the blood vessels and the heart is called cardiovascular disease (CVD) [1]. In the 2017 report of the American Heart Association, CVD is common and the top risk factor of mortality all over the world. In 2015, CVDs caused the death of 17.7 million individuals, with 6.7 million fatalities attributed to coronary artery disease (CAD). The most prevalent cardiovascular defect appears to be CAD; heart disease is a major cause of fatalities all over the world. According to research, one-third of all women, regardless of ethnicity or race, die from CAD. CAD, Alzheimer disease, bronchus, road injury, diarrheal diseases, diabetes mellitus, tuberculosis, stroke, lower respiratory infections, and lung cancers are among the top ten causes of death, according to the World Health Organization (WHO) [2]. Despite the fact that heart disease care has altered dramatically in recent decades, people with stable CAD are nonetheless at risk for a serious cardiovascular event. According to one study [3], in the United States, CAD strikes around 6% of the elderly. As a result, CAD is the most common of the CVDs. Furthermore, by 2030, the number of deaths is predicted to reach more than 22 million. As a result, early detection and medical attention will lower the number of CAD-related fatalities.

Atherosclerosis – the buildup of fatty deposits and cholesterol in the internal layers of the coronary artery – causes CAD [4]. The evolution of CAD is visually represented in Figure 13.1. In the absence of impediment, oxygen-rich blood flows freely via a normal coronary artery. Endothelial dysfunction develops as plaque deposits in the inner layer of the coronary artery.

As plaque thickness grows, the lumen of the coronary artery narrows, interrupting blood flow. Fatty streaks and constant plaque alterations in the coronary artery wall are the two types of modifications. The coronary plaque blockage will limit blood flow over time; as a result, the heart muscle receives insufficient oxygen-rich blood. [5] Furthermore, the lipid-rich plaque may become susceptible to rupture and may become unstable. Acute plaque rupture, which results in the blood clot creation on the plaque's facet that fully obstructs the artery lumen, is the most common reason for acute myocardial infarction (heart attack).

The injured and weakened heart muscle is the root cause of issues such as cardiac failure and arrhythmias. Many persons with CAD are asymptomatic until the illness has progressed to the point where they experience chest discomfort and shortness of breath. As a result, health screening is critical before the disease develops to an irreversible stage.

Analysis of Coronary Artery Disease

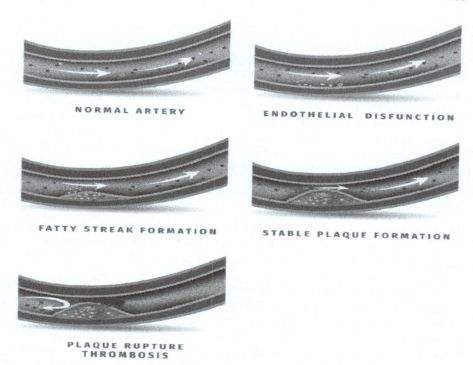

FIGURE 13.1 Progression of atherosclerosis and the genesis of CAD [4].

Traditional invasive coronary angiography has been established as the benchmark of diagnosing individuals with known or suspected CAD [6]. This method, however, is time-consuming, intrusive, and costly. Because this treatment normally requires a brief stay in the hospital, its invasiveness may cause some patients discomfort [7]. Furthermore, this method has a low but significant complication rate. The morphological examination of cardiac structures has been made possible thanks to electron-beam computed tomography (EBCT). This is due to EBCT's excellent temporal resolution and its use of prospective electrocardiographic triggering. The EBCT technique, however, wasn't regarded as the suitable method for detecting the existence of coronary stenosis due to its low spatial resolution. The development of computed tomography (CT) angiography improved the detection and treatment of CAD and also the measurement of heart function in various situations [8].

Machine learning (ML) techniques are now utilized in a variety of products. These technologies have been employed in medical applications for decades, and they have resulted in numerous advancements on various levels. Specifically, numerous ML algorithms for diagnosing and understanding illness progression have been proposed. To diagnose CAD, ML models, [9] namely decision trees, support vector machine (SVM), XG-Boost, neural network, naive Bayes, and random forest, have been used.

In this study, we have proposed seven ML classifiers to classify CAD and analyse the classification results. Prediction of CAD followed the process of data composition, where data were lifted from the Z-Alizadeh Sani dataset. Then, data preprocessing was done for normalizing the data. In data splitting, the complete dataset was divided into testing and training. Next, various model-based training data were trained. Then, in the classification step, validation of testing data was done by the trained data. Finally, different ML models used above were compared in terms of finding statistical performance, accuracy, specificity, and sensitivity.

The rest of the report is organized as follows: Section 13.2 represents the literature survey of various studies performed on the CAD dataset. Section 13.3 shows the methodology for classification of the CAD dataset. After that, results analysis is done in Section 13.4. Section 13.5 shows the discussion, and Section 13.6 represents the conclusion and future work of the paper.

13.2 LITERATURE SURVEY

Many studies on the diagnosis of CAD using various datasets and ML techniques have been conducted in the past few years. The Z-Alizadeh Sani dataset is the most up-to-date in the field of heart disease research. The Z-Alizadeh Sani dataset [10] is examined for this purpose.

A study by Alizadeh Sani et al. [11] supported the use of ECG data mining methods to identify CAD. To categorise the CAD, they employed the sequential minimum optimization (SMO) approach for feature optimization, followed by naive Bayes algorithms. Finally, employing the ten-fold cross-validation approach on the SMO-naive Bayes hybrid algorithms produced an accuracy of 88.52%, whereas the accuracy achieved by SMO was 86.95%, and naive Bayes achieved an accuracy of 87.22%.

For the detection of CAD, Hosseini et al. [12] developed the classification algorithms naive Bayes, SMO, neural networks, and bagging with SMO. In addition to gaining knowledge and confidence in CAD, positive characteristics have also been assessed. This led to the SMO algorithms achieving the maximum ten-fold cross-validation accuracy of 94%.

Using CI methods, Zanguei et al. [13] were able to identify three distinct types of severe coronary stenosis. Using analytic methods, they examined the importance of vascular stenosis features. In the end, the accuracy of the LAD, LCX, and RCA was 86.14%, 83.17%, and 83.50% using the SVM with ten-fold cross-validation. This was achieved using SVM. For CAD detection, Arabasdi et al. [14] proposed a neural network-genetic hybrid approach. Researchers used both genetic and neural network algorithms to evaluate the dataset, with the neural network approach showing an accuracy of 84.62% and the neural network-genetic algorithm yielding an accuracy of 93.85% when utilising the ten-fold cross-validation technique.

Khosravi et al. [15] employed a feature engineering technique for non-invasive CAD detection using naive Bayes, C4.52, and SVM [15]. They now have 500 samples instead of the 303 records they had previously. The accuracy of the naive Bayes, C4.52, and SVM algorithms was 86%, 89.8%, and 96%, respectively, when employing the ten-fold cross-validation approach. Nu-SVM, also known as N2 Genetic-NuSVM, was used by Abdar et al. [16] in their research. Using a two-level genetic method, it is being utilised to enhance the SVM parameters and select the feature at the same time. Using a ten-fold cross-validation approach, they found a 93.087% accuracy rate for CAD diagnosis (Table 13.1).

TABLE 13.1
A review table presenting the performance of different ML classifiers based on the Z-Alizadeh Sani dataset

Sl. No.	Authors & Year	Feature Extraction Method and Classifiers	Performance (Accuracy)	Use Case
1.	Alizadehsani et al. [11]	SMO, naive Bayes	88.52%	Based on symptoms and ECG features, diagnosis of CAD using data mining techniques.
2.	Alizadehsani et al. [12]	Naïve Bayes, SMO, neural networks and bagging with SMO	94.08%	For the diagnosis of CAD, data mining approach.
3.	Alizadehsani et al. [13]	Ten-fold cross-validation SVM model with feature selection	86.14%	Using computational intelligence methods for CAD detection.
4.	Arabasadi et al. [14]	Neural network-genetic algorithm	93.85%	Heart disease diagnosis is assisted by computer-aided decision making using hybrid neural network techniques.
5.	Alizadehsani et al. [15]	Naive Bayes, SVM algorithms, and C4.5	96.40%,	It is possible to diagnose heart disease in high-risk persons by the use of coronary artery stenosis predictions.
6.	Abdar et al. [16]	Two-level hybrid genetic algorithm	93.08%	Two-level hybrid genetic algorithms for diagnosis of CAD.
7.	Dipto et al. [17]	SVM, logistic regression and ANN	93.3%	Prediction of CAD with different ML algorithms.
8.	Hassannataj Joloudari et al. [9]	RT, SVM, CHAID, decision tree	91.47%	CAD diagnosis; ranking significant features using RTs.
9.	Alizadehsani et al. [18]	SMO, naïve Bayes, C4.5, and Ada-Boost	82%	Diagnosis of CAD using data mining.
10.	Sharma et al. [19]	PCA, firefly optimization, decision tree	93.3%	Efficient predictive modelling for the classification of CAD.
11.	Abdar et al. [20]	Ensemble learning techniques	94.66%	An ensemble clinical decision support system for CAD diagnosis with nested ensembles.
12.	Cüvitoğlu et al. [21]	PCA, SVM, NB, ANN	85.55%	Principal component analysis & ML methods for classification of CAD datasets.

13.3 MATERIAL AND METHODS

13.3.1 Dataset Description

According to Alizadeh et al. [10], a dataset referred to as Z-Alizadeh Sani dataset is employed in this study. One of the most well-known datasets for automatic CAD recognition in ML is this one. As you can see, there are 303 patients in this database; many of them have CAD. The patient is classified as having CAD if one or more of their coronary arteries are stenotic. When the diameter of a coronary artery narrows by 50% or more, it is said to be stenotic. As a result, 216 patients had CAD, and the dataset includes 88 patients who were in a normal state on the Z-Alizadeh Sani. Each row has 55 features that can be utilized to determine if a patient has CAD or not.

13.4 METHODOLOGY

13.4.1 Data Normalization

After data collection, preparation is necessary for the analysis process. As a result, the data were both numerical and stringent. First, the number of records is converted from nominal to numerical values, such as gender, cerebrovascular accident, airway illness, chronic renal failure, thyroid disease, dyslipidemia, congestive heart failure, etc. We normalize all data before preprocessing by converting them to a numeric value between 0 and 1. It was necessary to transform the string to numeric data after the numbers had been normalized. As a result, a number between [0,1] was assigned to the string data. Sex is only one example of a variable that can have either male or female values of 0 or 1. In general, normalization leads to an increase in the accuracy of the classification models (Figure 13.2).

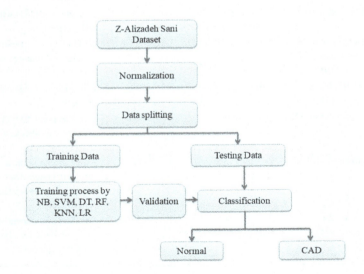

FIGURE 13.2 Flow chart of the proposed methods.

Analysis of Coronary Artery Disease

13.4.1.1 Data Splitting

To examine the process, the splitting phase is employed to create the training and testing data. Our entire database is divided into testing and training data, with 20% of the data being used for testing purposes and 80% being used for training purposes.

13.4.1.2 Classification Models

Classification relies on ML models to analyse data into two distinct sets, one for training and one for testing. Data were utilised to train KNN, decision tree, SVM, random forest XG-Boost, logistic regression, and naive Bayes ML models. The testing data were then confirmed using training data that have a higher classification accuracy rate. The following is a detailed description of seven different algorithms.

13.4.1.3 Support Vector Machine

With the help of non-straight planning, the SVM algorithm transforms the initially prepared data into a higher-dimensional space [22], where it searches for the linearly optimal separating hyperplane that separates the class tuples from one another. In order to separate data, a hyperplane with an adequate nonlinear mapping to a high-enough dimension is always an option. To find a hyperplane, the algorithm relies on support vectors and margins (Figure 13.3).

The decision function is applicable to data samples that may be separated linearly. When the data samples are not linearly separable, the resulting function is as follows:

$$y(x) = sgn\left[\sum_{i=1}^{m} \alpha_i y_i K(x_i, x) + b\right] \tag{13.1}$$

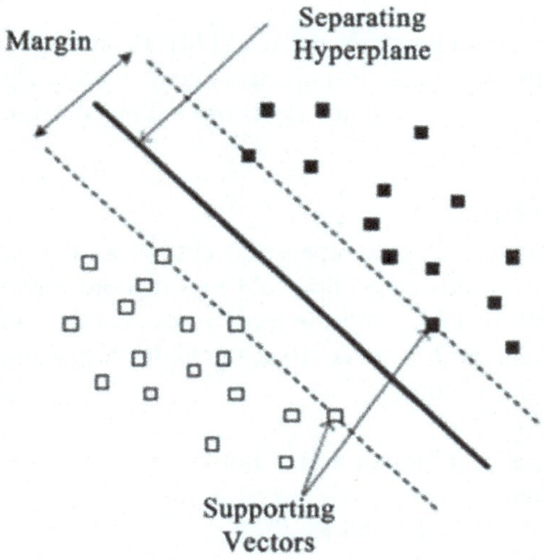

FIGURE 13.3 Support vector machine classifier.

where $K(x_i, x) = (\varphi(x1), \varphi(x))$ and $\varphi(x)$ is the non-linear space from the original space to high-dimensional space. All the basic kernel is given below [23]:

- Linear: $K(x_i, x_j) = x_i^T x_j$
- Polynomial: $K(x_i, x_j) = (\gamma x_i^T x_j + r)^d \gamma > 0$
- Radial Bias Function (RBF): $K(x_i, x_j) = \exp(-\gamma \|x_i - x_j\|^2), \gamma > 0$
- Sigmoid: $K(x_i, x_j) = \tanh(\gamma x_i^T x_j + r)$

13.4.1.4 Decision Tree

Decision tree is a technique that is used to work with both category and numerical data. Tree-like structures are created using decision trees. Medical datasets are frequently handled using decision trees. The data in tree-shaped graphs are simple to implement and analyse. The study is based on three nodes in the decision tree model.

- Root node: It is the most important node, and it serves as the foundation for all other nodes.
- Interior node: It is in charge of numerous properties.
- Leaf node: This node indicates the outcome of each test.

The data are divided into two or more equivalent sets by this algorithm. The entropy of each characteristic is calculated, and data are divided into predictors with most info gain or the least entropy:

$$Entropy(S) = \sum_{i=1}^{c} -Pi \log_2 Pi \tag{13.2}$$

$$Gain(S, A) = Entropy(S) - \sum_{v \in Values(A)} \frac{|Sv|}{|S|} Entropy(Sv) \tag{13.3}$$

The results are simple to understand and read [24]. Because it analyses the dataset in a tree-like graph, this technique is more accurate than other algorithms. However, the data may be reclassified, and for decision-making purposes, only one characteristic is checked at a time.

13.4.1.5 Random Forest

The random forest algorithm is a supervised classification method. The forest is created by a few trees in this algorithm. [25] In a random forest, each tree emits a class assumption, and the class with the most votes becomes a models conjecture. The more trees in a random forest classifier, the higher the accuracy. The following are the three standard procedures:

- Forest RI (Random information decision);
- Forest RC (Random mix);
- Combination of Forest RI and RC.

It excels at classification tasks and can overcome missing characteristics. Furthermore, it takes a long time to get predictions because it needs more trees and massive data; therefore, results are unreachable.

13.4.1.6 K-Nearest Neighbor (K-NN)

The K-NN algorithm is a type of classification algorithm. It describes objects that are reliant on their immediate surroundings. It's a type of event-based learning. The distance between a characteristic and its neighbors is calculated using the Euclidean distance [26]. It takes a collection of focuses that are named and then utilises it to stamp other points in the most efficient way possible. The data are grouped together according to their resemblance, and K-NN can be used to fill in the gaps in the data. When the missing qualities are filled in, the data set is subjected to various prediction strategies. Using different combinations of these algorithms, it is possible to improve accuracy. Without a model or different suppositions, it is simple to complete the K-NN algorithm. This algorithm can be used for regression & classification, or searching. Despite the fact that K-NN is the most straightforward algorithm, uproarious and insignificant highlights influence its accuracy.

13.4.2 LOGISTIC REGRESSION

Logistic regression is a classification algorithm that predicts the likelihood of a variable [27]. One or zero are the only possible values for the variable. $P(X = 1)$ is predicted numerically by a logistic regression model. Different classification difficulties, such as spam identification and diabetes prediction, can be addressed using one of the most straightforward ML algorithms.

13.4.3 TYPES OF LOGISTIC REGRESSION

In general, logistic regression implies binary logistic regression with binary objective factors, but it can also predict two additional classes of target factors [17,28]. Due to the obvious vast number of classes, logistic regression can be classified as follows.

- Binary or binomial
- Multinomial
- Ordinal

13.4.3.1 Logistic Regression Assumptions

- We should be aware of the logistic regression's accompanying assumptions.
- The objective factors in binary logistic regression should always be binary, and the ideal result is addressed by the variable level 1.
- The model should not have any multi-collinearity, which means the autonomous factors should be independent of one another.

- Important factors for our model should be remembered.
- For logistic regression, we should choose a large sample size.

13.4.3.2 Naïve Bayes

The naive Bayes classifier is a supervised algorithm. The Bayes hypothesis is used in a simple classification procedure. It expects ascribes to have strong (naive) autonomy. The Bayes hypothesis is a mathematical formula for calculating probability. The predictors are not linked to one another or identified with one another. Every one of the properties adds to the probability of amplifying it on its own [28].

$$P\left(\frac{X}{Y}\right) = \frac{P\left(\frac{Y}{X}\right) * P(X)}{P(Y)} \tag{13.4}$$

The back probability is P(X/Y), the class earlier probability is P(X), the predictor earlier probability is P(Y), and the probability is P(Y/X). Naive Bayes is a simple, straightforward, and effective classification algorithm for non-linear, jumbled data. Regardless, there is a lack of precision because it is based on speculation and class restrictive freedom.

13.4.3.3 XG-Boost

The XG-Boost is an implementation of gradient boost decision trees that is made for better speed and performance. The model's implementation includes the features of the scikit-learn and R implementations, as well as novel enhancements like regularization.

The gradient boosting decision tree technique is implemented in the XG-Boost library.

13.5 RESULT ANALYSIS

A performance matrix is used to depict performance of any classifier by calculating assessment parameters [25]. Accuracy: Efficiency of any algorithm can be evaluated on the accuracy of CAD forecast, which is given by the Equation 13.5.

$$Accuracy = \frac{TP + TN}{TP + FP + FN + TN} \tag{13.5}$$

Sensitivity may be calculated mathematically as follows:

$$Sensitivity = \frac{TP}{TP + FN} \tag{13.6}$$

The greater the sensitivity, the fewer false negatives and the greater the value of true positives.

Specificity may be calculated mathematically as follows:

Analysis of Coronary Artery Disease 197

$$Specificity = \frac{TN}{TN + FP} \qquad (13.7)$$

where TN = True Negative, TN = True Negative and FP = False Positive.

The greater the specificity, the lower the level of false positives and the greater the value of genuine negatives. The lower the specificity number, the greater the false positive value and the lower the real negative value, ROC and AUC curve [29]. TPR and FPR are the two data that determine the ROC curve.

13.5.1 Performance Comparison of Algorithms

This section displays the classified results from several prediction models. We compared several models using various parameters, which were accuracy, sensitivity, and specificity. The comparison with five models is shown in Table 13.2. This compares several ML algorithms for analyzing CAD based on accuracy, sensitivity, and specificity.

As shown in Table 13.2, the results of training the models on the dataset demonstrate that random forest fared better than the other five techniques, with the greatest accuracy of 93.44%. That's also obvious from the ROC graph plotted for all of the methods trained on the dataset.

In SVM, the hyperplane is determined by one of multiple kernels. We have experimented with four kernels: linear, poly, RBF, and sigmoid. (Figure 13.4).

The linear kernel fared the finest for this dataset, as seen in the figure above, with a score of 91.8%~92%.

In KNN, the number of neighbors can be varied. So, we varied the number of neighbors from 1–40 and computed the score of the test in every case. Therefore, in each scenario, we produced a line graph of the number of neighbors and the test score obtained (Figure 13.5).

TABLE 13.2
Performance comparison of machine learning algorithms using CAD dataset

Classification Algorithm	Accuracy (with Ten-Fold)(%)	Accuracy (without Ten-Fold)(%)	Sensitivity (%)	Specificity (%)
SVM-linear	85.46	91.8	82.3	95.4
Decision Tree	76.92	91.8	88.24	93.18
Logistic Regression	83.85	85.25	64.71	93.18
KNN	71.28	73.77	5.88	100
Naive Bayes	52.43	55.74	88.24	95.45
XGBoost	85.82	83.61	64.71	90.91
Random Forest	**85.54**	**93.44**	**76.47**	**100**

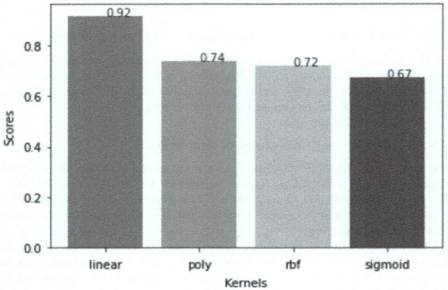

FIGURE 13.4 SVM score for various kernels.

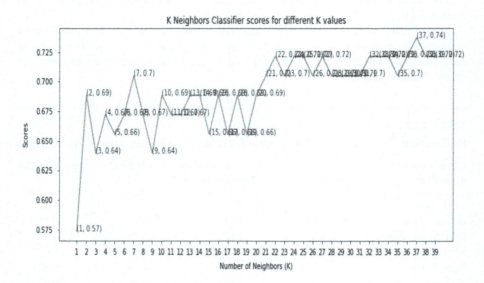

FIGURE 13.5 KNN scores for various K values.

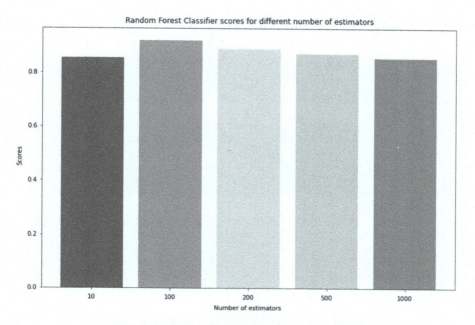

FIGURE 13.6 Random forest scores for various estimator's value.

As you've seen, when the number of neighbors was set to 37, we received the highest possible score of 73.77%~74%.

Random forest generates a forest of trees, with each tree produced by a randomized feature extraction from the whole set of features. We changed the number of trees used to forecast the class. We computed test scores for 10, 100, 200, 500, and 1000 trees. Following that, we graphed these scores on a bar graph to discover which produced the greatest outcomes (Figure 13.6).

Looking just at thecbar graph, we can observe that the highest score was obtained for 100 trees (Figures 13.7–13.14).

13.6 DISCUSSION

Several researchers have used other methods before working on the Alizadeh Sani dataset for evaluation of detection and classification of CAD. We compare our proposed method to other methods that used the Alizadeh Sani dataset, in prior studies. Table 13.3 compares the accuracy, sensitivity, and specificity of several methodologies used in prior research.

In this study, we investigated and examined several ML algorithms for predicting CAD, including SVM, decision tree, KNN, random forest, XGBoost, logistic regression, and naive Bayes. Prediction of CAD pursued the step of data preparation in which details were taken from the Z-Alizadeh Sani dataset. Then, data preprocessing was done to normalize the data. Data splitting was done, in which the entire data were split into testing and training.

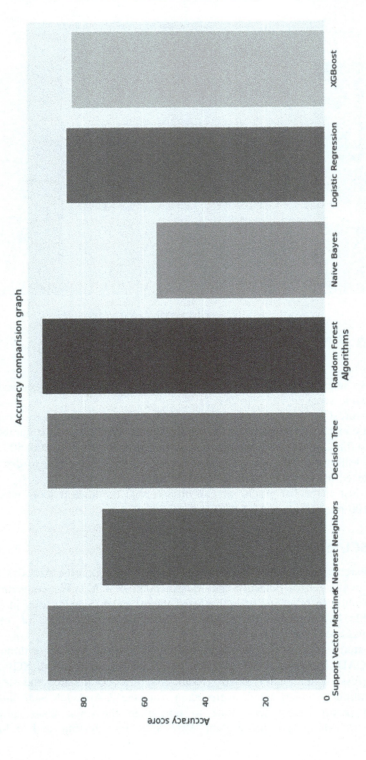

FIGURE 13.7 Accuracy comparison bar graph.

Analysis of Coronary Artery Disease

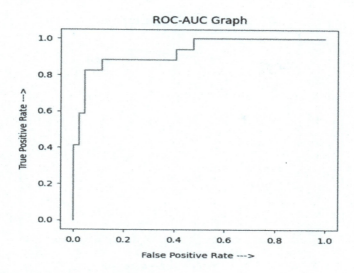

FIGURE 13.8 ROC-AUC graph of SVM.

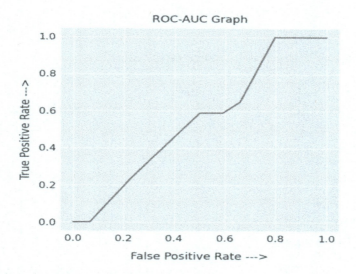

FIGURE 13.9 ROC-AUC graph of KNN.

Eventually, various model-based training data were trained, and testing data were validated by the trained data in the step of classification. We assessed the sensitivity, specificity, and accuracy and discovered that a random tree classifier produced the best results, with 93.44% accuracy, 76.47% sensitivity, and 100% specificity.

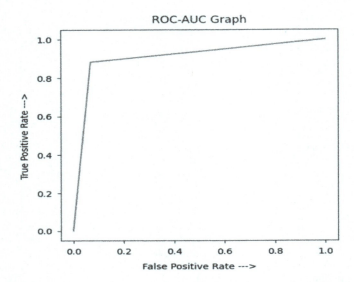

FIGURE 13.10 ROC-AUC graph of decision tree.

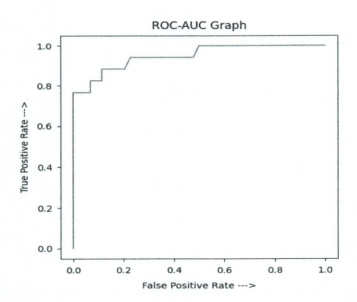

FIGURE 13.11 ROC-AUC graph of random forest.

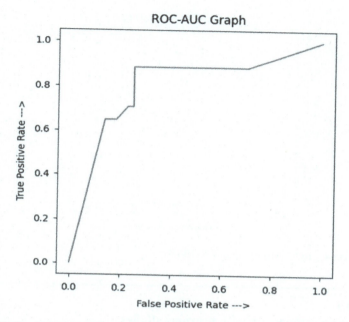

FIGURE 13.12 ROC-AUC graph of decision tree.

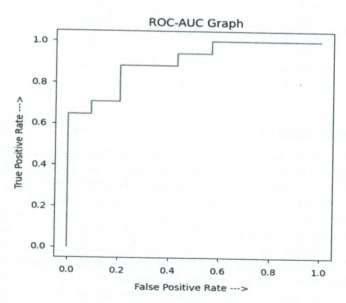

FIGURE 13.13 ROC-AUC graph of random forest.

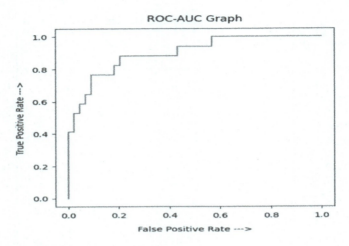

FIGURE 13.14 ROC-AUC graph of XGBoost.

TABLE 13.3
Performance comparison with previous studies

Reference	Methods	Accuracy (%)	Sensitivity (%)	Specificity (%)
[30,31]	ANN	85.85	72.50	91.21
[21]	SVM	86.14	90.96	79.37
[26]	Random Tree	91.47	–	–
[27]	SMO	82.16	90.74	60.92
Our study	Proposed	93.44	76.47	100

13.7 CONCLUSION

Numerous methods were used on the ZAlizadeh Sani dataset in this research work, and the findings were discussed. According to our medical understanding, the traits in this dataset might be indicative of CAD. Furthermore, the characteristics employed in this report may be tested along with little side effects and expenditures. As a result, using the suggested method may detect the CAD condition with cheap cost and great probability. In the future, we hope to examine forecasting the status of each artery separately. Furthermore, it is evident that accurate diagnosis of unhealthy individuals is more crucial than accurate detection of healthy individuals. As a result, an additional objective to achieve is to use budget conscious algorithms to take this element into account. Eventually, bigger datasets with more characteristics, as well as wider data mining methodologies, might be employed to provide bigger and more important findings.

REFERENCES

[1] Fujita, H., Sudarshan, V. K., Adam, M., Oh, S. L., Tan, J. H., Hagiwara, Y., ... & Acharya, U. R. (2017, June). Characterization of cardiovascular diseases using wavelet packet decomposition and nonlinear measures of electrocardiogram signal. In *International Conference on Industrial, Engineering and Other Applications of Applied Intelligent Systems* (pp. 259–266). Springer, Cham.

[2] Chandrakar, R., Raja, R., Miri, R., Sinha, U., Kushwaha, A. K. S., & Raja, H. (2022). Enhanced the moving object detection and object tracking for traffic surveillance using RBF-FDLNN and CBF algorithm, *Expert Systems with Applications*, 191, 116306, ISSN 0957-4174, 10.1016/j.eswa.2021.116306

[3] Mozaffarian, D., Benjamin, E. J., Go, A. S., Arnett, D. K., Blaha, M. J., Cushman, M., ... Turner, M. B. (2016). Heart disease and stroke statistics—2016 update: A report from the American Heart Association. *Circulation*, 133(4), e38–e360.

[4] Pandey, S., Miri, R., Sinha, G. R., & Raja, R. (2022). AFD filter and E2N2 classifier for improving visualization of crop image and crop classification in remote sensing image. *International Journal of Remote Sensing*, 43(1), 1–26. 10.1080/01431161.2021.2000062

[5] Acharya, U. R., Fujita, H., Lih, O. S., Adam, M., Tan, J. H., & Chua, C. K. (2017). Automated detection of CAD using different durations of ECG segments with convolutional neural network. *Knowledge-Based Systems*, 132, 62–71.

[6] Chandrakar, R., Raja, R., Miri, R., Patra, R. K., & Sinha, U. (2021). Computer succored vaticination of multi-object detection and histogram enhancement in low vision. *Int. J. of Biometrics. Special Issue: Investigation of Robustness in Image Enhancement and Preprocessing Techniques for Biometrics and Computer Vision Applications*, 3(1), 1–12.

[7] Sun, Z. (2013). *Coronary Computed Tomography Angiography in CAD: A Systematic Review of Image Quality, Diagnostic Accuracy and Radiation Dose.* Nova Science Publishers, Incorporated.

[8] Sun, Z., Choo, G. H., & Ng, K. H. (2012). Coronary CT angiography: Current status and continuing challenges. *The British Journal of Radiology*, 85(1013), 495–510.

[9] Hassannataj Joloudari, J., Hassannataj Joloudari, E., Saadatfar, H., GhasemiGol, M., Razavi, S. M., Mosavi, A., ... Nadai, L. (2020). CAD Diagnosis; Ranking the Significant features using random trees model. arXiv e-prints, arXiv-2001.

[10] Tiwari, L., Raja, R., Awasthi, V., Miri, R., Sinha, G. R., & Alkinani, M. H. (2021). Polat, detection of lung nodule and cancer using novel Mask-3 FCM and TWEDLNN algorithms, *Measurement*, 172, 108882, ISSN 0263-2241. 10.1016/j.measurement.2020.108882

[11] Alizadehsani, R., Habibi, J., Hosseini, M. J., Boghrati, R., Ghandeharioun, A., Bahadorian, B., & Sani, Z. A. (2012). Diagnosis of CAD using data mining techniques based on symptoms and ECG features. *European Journal of Scientific Research*, 82(4), 542–553.

[12] Alizadehsani, R., Habibi, J., Hosseini, M. J., Mashayekhi, H., Boghrati, R., Ghandeharioun, A., Bahadorian, B., & Sani, Z. A. (2013). A data mining approach for diagnosis of CAD. *Computer Methods and Programs in Biomedicine*, 111(1), 52–61.

[13] Alizadehsani, R., Zangooei, M. H., Hosseini, M. J., Habibi, J., Khosravi, A., Roshanzamir, M., Khozeimeh, F., Sarrafzadegan, N., & Nahavandi, S. (2016). CAD detection using computational intelligence methods. *Knowledge-Based Systems*, 109, 187–197.

[14] Arabasadi, Z., Alizadehsani, R., Roshanzamir, M., Moosaei, H., & Yarifard, A. A. (2017). Computer aided decision making for heart disease detection using hybrid neural network-Genetic algo. *Computer Methods and Programs in Biomedicine*, 141, 19–26.

[15] Alizadehsani, R., Hosseini, M. J., Khosravi, A., Khozeimeh, F., Roshanzamir, M., Sarrafzadegan, N., & Nahavandi, S. (2018). Non-invasive detection of CAD in high-risk patients based on the stenosis prediction of separate coronary arteries. *Computer Methods and Programs in Biomedicine*, 162, 119–127.

[16] Abdar, M., Książek, W., Acharya, U. R., Tan, R. S., Makarenkov, V., & Pławiak, P. (2019). A new ML technique for an accurate diagnosis of CAD. *Computer Methods and Programs in Biomedicine*, 179, 104992.

[17] Dipto, I. C., Islam, T., Rahman, H. M. M., & Rahman, M. A. (2020). Comparison of different ML algos for the prediction of CAD. *Journal of Data Analysis and Information Processing*, 8, 41–68. 10.4236/jdaip.2020.82003

[18] Alizadehsani, R., Habibi, J., Sani, Z. A., Mashayekhi, H., Boghrati, R., Ghandeharioun, A., & Bahadorian, B. (2012). Diagnosis of CAD using data mining based on lab data and echo features. *Journal of Medical and Bioengineering*, 1(1).

[19] Sharma, G., Rani, G., & Dhaka, V. S. (2021, March). Efficient predictive modelling for classification of CADs using ML approach. *IOP Conference Series: Materials Science and Engineering*, 1099(1), p. 012068. IOP Publishing.

[20] Abdar, M., Acharya, U. R., Sarrafzadegan, N., & Makarenkov, V. (2019). NE-nu-SVC: A new nested ensemble clinical decision support system for effective diagnosis of CAD. *IEEE Access*, 7, 167605–167620.

[21] Cüvitoğlu, A., & Işik, Z. (2018, May). Classification of CAD dataset by using principal component analysis and ML approaches. In *2018 5th International Conference on Electrical and Electronic Engineering (ICEEE)* (pp. 340–343). IEEE.

[22] Gudadhe, M., Wankhade, K., & Dongre, S. (2010). Decision support system for heart disease based on support vector machine and artificial neural network. In *IEEE International Conference on Computer and Communication Technology*, Allahabad, 17 September, 2010 (pp. 741–745). 10.1109/ICCCT.2010.5640377

[23] Hsu, C. W., Chang, C. C., & Lin, C. J. (2003). A practical guide to support vector classification. Department of Computer Science National Taiwan University, Taipei 106, Taiwan, http://www.csie.ntu.edu.tw/~cjlin Initial version: 2003 Last updated: May 19, 2016.

[24] Ghiasi, M. M., Zendehboudi, S., & Mohsenipour, Ali A. (2020). Decision Tree-based diagnosis of CAD: CART Model, *Computer Methods and Programs in Biomedicine*, 192, p. 105400. 10.1016/j.cmpb.2020.105400

[25] Alotaibi, S. S., Almajid, Y. A., Alsahali, S. F., Asalam, N., Alotaibi, M. D., Ullah, I., & Altabee, R. M. (2020, October). Automated prediction of CAD using Random Forest and Naïve Bayes. In *2020 International Conference on Advanced Computer Science and Information Systems (ICACSIS)* (pp. 109–114). IEEE.

[26] Kumar, S., Raja, R., & Gandham, A. (2020). Tracking an object using traditional MS (Mean Shift) and CBWH MS (Mean Shift) algorithm with Kalman filter. In: Johri P., Verma J., Paul S. (eds.), *Applications of Machine Learning. Algorithms for Intelligent Systems*. Springer, Singapore. pp. 47–65. 10.1007/978-981-15-3357-0_4

[27] Kurt, I., Ture, M., & Kurum, A. T. (2008). Comparing performances of logistic regression, classification and regression tree, and neural networks for predicting CAD. *Expert Systems with Applications*, 34(1), 366–374.

[28] R. Raja, S. Kumar, S. Choudhary, & H. Dalmia (2021). An Effective Contour Detection based Image Retrieval using Multi-Fusion Method and Neural Network, *Submitted to Wireless Personal Communication, PREPRINT (Version 2)* available at Research Square. 10.21203/rs.3.rs-458104/v1

[29] Parikh, R., Mathai, A., Parikh, S., Sekhar, G. C., & Thomas, R. (2008). Understanding and using sensitivity, specificity and predictive values. *Indian Journal of Ophthalmology*, 56(1), 45.

[30] Raja R., Patra R. k., & Sinha T. S. (2017). Extraction of features from dummy face for improving biometrical authentication of human. *International Journal of Luminescence and Application*, 7(3–4), October–December 2017, Article 259, 507–512, ISSN:1 2277-6362.

[31] Chandrakar, R., Raja, R., & Miri, R. (2021). Animal detection based on deep convolutional neural networks with genetic segmentation. *Multimed Tools and Applications*.pp. 1–14. 10.1007/s11042-021-11290-4

Index

advantages of cloud computation, 142
Amazon Elastic Block Store (EBS volumes), 144
Amazon Elastic Compute Cloud, 143
Amazon Elastic File System (Amazon EFS), 144
Amazon Machine Learning, 145
Amazon Web Services, 143
analysis of deep learning techniques, 21
application areas, 44
application in agriculture using color classification, 45
application in color constancy, 44
artificial neural network, 3
authentication phase, 164
AWS Elastic Beanstalk, 142
AWS – Healthcare Solutions, 147

bidirectional long short term memory, 47
big data analysis in AWS, 145
binary jaya algorithm, 9
body temperature, 76

chatbot using NLP, 130
classification of WSNs protocols, 152
clinical trials, 127
clone attack, 159
cloud computing models, 140
cloud healthcare management service, 141
cloud pricing strategy, 145
clustering phase, 154
color blindness detection, 44
common EEG artifacts, 176
computational intelligence techniques, 2
conditional GAN, 45
convolution neural network, 22, 45
convolution neural network experiment design and results, 24
Coronary Artery Disease (CAD), 188

data collection, 57
data from sensors, 81
data inconsistency, inaccuracy, and missing values, 5
data normalization, 192
data pre-processing, 176
data preprocessing, 108
data splitting, 193
data-centric routing protocol, 152
database description – BONN University EEG dataset, 174

dataset, 19, 32
decision tree, 180, 194
decryption and authentication, 165
deep learning and convolution neural network, 30
deep learning techniques, 16, 108
deep learning techniques, 16
deep-learning combined with SVM approach, 90
denial of service attack, 156
DenseNet, 50, 111
deployment models, 140
diagnosing method for oral cancer, 87
Discrete Energy Separation Algorithm (DESA), 56
drug discovery, 124

efficient clustering of modules to increase safety of WSN, 162
electroencephalogram (EEG), 170
emotion detection, 76
ethical and privacy issues, 6
evolutionary algorithms, 3
extreme learning machine, 179
eye blink detection, 75

face detection, 74
face recognition system, 73
feature classification by Deep Neural Network (DNN) classifier, 61
feature engineering, 6
filter method, 6
food and breed hunting for animals, 44
framework of the proposed techniques, 58
fusion of the results obtained from transfer learning and SVM process, 95
future development, 133
fuzzy system, 4

Gaussian Mixture Model (GMM), 56
gene expression dataset, 8

hardware and sensors used, 77
health insurance and fraud detection, 130
health parameters and stress detection, 75
heart rate, 77
hierarchical routing protocol, 153
histogram analysis of network model using LSTM, 66
imbalanced data, 5
Infrastructure as a Service (IaaS), 140

issues and challenges, 4

K-nearest neighbors, 179

logistic regression, 180
Long Short Term Memory (LSTM) technique, 62

medical diagnosis, 131
medical image and diagnostics, 126
medical imagery/brain tumor detection, 31
multiple path routing protocol, 153

network architecture, 110

patient monitoring and personalized treatment, 128
performance comparison of algorithms, 197
physiological parameters and sensors description, 76
Platform as a Service (PaaS), 140
preprocessing of data and feature extraction using Wavelet Packet Decomposition (WPD), 57
privacy and security issues in WSN, 155
probabilistic neural network, 48

progression of Alzheimer's disease, 106
proposed solution, 73
PyEEg framework, 177

quality of service-based routing protocol, 155

random forests, 179
Rapid Eye Movement (REM), 56
Rectified Linear Unit (ReLU), 24
recurrent neural network, 22

sentiment analysis based on color attributes, 45
sinkhole attack, 158
Software as a Service (SaaS), 140
Support Vector Machines (SVMS), 3, 178
systems, 72

validation and testing accuracy, 38
VGG-16, 49

white blood cell classification, 17
Wireless Sensor Network (WSN), 150

XG-boost, 196